THE PALYNOLOGY OF ARCHAEOLOGICAL SITES

Geoffrey W. Dimbleby
Emeritus Professor
Institute of Archaeology
University of London
London, England

1985

ACADEMIC PRESS, INC.
(Harcourt Brace Jovanovich, Publishers)
London Orlando San Diego New York
Toronto Montreal Sydney Tokyo

ACADEMIC PRESS INC. (LONDON) LTD.
24–28 Oval Road
LONDON NW1 7DX

United States Edition published by
ACADEMIC PRESS, INC.
Orlando, Florida 32887

British Library Cataloguing in Publication Data

Dimbleby, G. W.
The palynology of archaeological sites.

Includes bibliographical references and index.
1. Palynology. 2. Plant remains (Archaeology)
3. Excavations (Archaeology) I. Title.
CC79.5.P58D56 1984 930.1'028'3 84-9340
ISBN 0-12-216480-6 (alk. paper)

PRINTED IN THE UNITED STATES OF AMERICA

85 86 87 88 9 8 7 6 5 4 3 2 1

CONTENTS

PREFACE

Pollen analysis now has a long tradition behind it, reflected in a large and still growing body of literature. It has been applied in many different circumstances, from the tropics to the poles, and on materials as different as lake deposits, peats, soils, gravels, and even ice. Conventionally, pollen analysis has sought to establish a chronosequence of vegetation in the Pleistocene, and most particularly in the postglacial period, and some pollen analysts would see this as its only legitimate function. But the presence of pollen in materials or deposits can sometimes be made to tell quite a different story, one more closely associated with man himself than with secular changes of climate or vegetation upon which man could have no influence. Since such information comes from the analysis of pollen from appropriate materials, such studies clearly fall within the scope of pollen analysis, or palynology, even if this means widening an originally narrow definition: it is not justifiable to restrict these terms merely to the original aim of establishing climatic change. In other words, pollen analysis as a tool need not be confined to its traditional use. From the archaeological point of view this point is important, for polliniferous deposits may occur which would be quite unsuitable as a basis for paleoecological studies, but which can give often unexpected information about human and animal activities.

Over the years, pollen analysis has shed much light on the relationships between man and his environment throughout his cultural history. Insofar as the bulk of this evidence comes from stratified waterlogged deposits, the traditional resource of the pollen analyst, its application to archaeology must be indirect. For obvious reasons, such deposits are seldom associated with human occupation. However, the valuable indirect contribution made to archaeology in this way has opened up the possibility that archaeological sites, including occupation sites, could themselves be sources of pollen which could be related directly to contemporary human activities. It is the purpose of this book to draw together such information, not only to show what has been done, but also to indicate to archaeologists the ways in which pollen might in future be used as a research tool. It will also be shown that not all materials which contain pollen can usefully be used for pollen analysis.

To this end the term *site* will not be defined too closely. It might indicate a farmstead, for instance, including not only the settlement but also the associated field system and boundary banks. It could well include such outworks even though the settlement site itself had not been found or had been destroyed. But the term *site* will not normally be applied to place where archaeological material is found in a secondary context; for example, the finding of rolled hand-axes in a gravel deposit. On the other hand, the finding of hand-axes in a primary setting would constitute a site. There are, of course, blurred edges: hillwash deposits may contain artifacts in a secondary context, but pollen analysis of the deposit itself can tell us something about how the deposit came to reach its new place of rest, a process in which man himself often played a part.

Nor will *site* be taken to include the many examples of archaeological material being found in peat bogs, unless it can be shown that there had been actual occupation of some sort at these points (see Chapter 2). In the Fenlands of East Anglia, the Somerset Levels, and the bogs of the Low Countries and Denmark, many cases are known of isolated artifacts found in stratified deposits in the peat, and these can give direct links between the culture and its environment, but for present purposes I am not including these as sites. I also exclude bog burials, through in one sense they could be said to represent occupation.

As said above, pollen analysis has been applied over a wide geographical range and for this reason no geographical restriction is applied in this book. Whilst it is generally true that pollen analysis has its greatest application in cool temperature regions, this is not exclusively so. Indeed, pollen may be preserved wherever microbiological action is inhibited. Such may be the case where the deposit is waterlogged, but other factors may produce the same inhibition. Extreme aridity can lead to the preservation of pollen in desert areas, whilst soil chemistry may produce similar effects. High acidity, even in aerated deposits, may inhibit microbiological breakdown of pollen, as may high levels of salt or some metals, particularly copper (see Chapter 1).

I owe my readers some apology for what may seem undue emphasis here on my own research over the years. I became involved with palynology quite incidentally, using sites as a means of establishing paleoecology in an area which had no peat deposits (see discussion in Chapter 3). Sampling an old land surface (OLS) beneath an earthwork often led to the sampling of the body of the earthwork itself, usually to provide a contrast to the OLS. The analyses of these extra samples often produced unexpected and interesting results which could be interpreted in terms of the structure or construction of the earthwork (see Chapter 4). Generally the only data published and discussed were those relating to the OLS, with the result that over the years a body of data has accumulated which has never been adequately evaluated and interpreted in archaeological terms. It seemed to me that if I did not take on such an evaluation it was unlikely that anyone else ever would. As my own observations were the initial reason for this approach, they clearly must feature largely in the overall treatment. I have tried to deal with them as unobtrusively as possible, and around this core I have included many aspects of site palynology with which I myself have had little or no connection (e.g., Chapters 2 and 6) in order to give a more balanced archaeological coverage.

In every environmental context there are a number of processes operating and interacting; microbiology is only one factor, and pollen is only one of a number of possible types of sub-fossils which may be preserved. Whilst much may be learnt from one such source, corroborative evidence from other sources adds to the confidence which can be placed on the primary source. Remains of insects, molluscs, or seeds, as well as the soil characters, may all contribute towards an environmental interpretation. At times, evidence from different sources may on the face of it appear to be contradictory. Ultimately, of course, there can be no contradiction, so discrepancy is to be seen as evidence of incompleteness; if pursued it may lead to a depth of information greater than could have been achieved by a single line of investigation (see Chapter 1). Discrepancy may also be found between the conclusions from pollen analyses of archaeological sites and from those stratified deposits not associated with sites (see Chapter 3). Neither approach is to be seen as more correct or more reliable than the other, but rather that the two reflect different but complementary sources of pollen, and perhaps different conditions of preservation. Archaeological sites are a fragment of the past which have survived to the present day. They may contain not only artifacts but also sub-fossil organic remains which equally are survivors from the past. These should be approached positively and not dismissively, and it is hoped that they can produce results which could not have been obtained from any other source.

It is an inherent weakness of a book dealing with one topic in this way that it concentrates on only one approach to the exclusion of others; in ecology this is particularly to be regretted. However, it is apparent that I can do no more than draw attention to the importance of a multidisciplinary approach wherever this is possible. At several points in this book I have indicated where this has been done, usually to advantage.

Amongst other possible approaches, of course, the archaeological one almost inevitably arises in this particular context. But just as an environmentalist must accept the validity of archaeological evidence, so must the archaeologist be prepared to accept the validity of the scientific evidence. Some archaeologists are glad to have pollen evidence if it supports their theories, but if it does not they feel free to dismiss it. I had one example, for which I will not quote chapter and verse, which illustrates this point. In excavating a round barrow on moorland, the excavator submitted samples for pollen analysis, one from the OLS and another from the surface beneath what he took to be the dump from a nineteenth century (A.D.) robber trench. Other work in the area suggests that the site had been under moorland for virtually the whole of the historic period, yet the sample from beneath the dump from the robber trench gave a pollen spectrum of mixed deciduous woodland of primary aspect; it seems impossible to equate this with the vegetation in the nineteenth century. Though the pollen data were published in an appendix, the excavator ignored this awkward piece of evidence in his report, leaving a complete contradiction for anyone who took the trouble to examine the results in detail. As in so many rescue excavations, there was no means of going back to check the point in the field, as the barrow was subsequently destroyed, but clearly the disparity meant that the separate archaeological and environmental approaches between them had not revealed the full truth about this small earthwork. No doubt the archaeological report was written before the post-excavation work was completed. In this case the pollen evidence was ignored; I could quote other cases in which the archaeologist responded to a similar situation by not publishing the conflicting evidence at all. It seems to be assumed that one should only publish interdisciplinary work if it successfully checks at all points. This raises in a different way the old problem in science—does one publish experimental results which do not bear out the hypothesis on which the experiment was based? Hidden amongst these may be a very large chunk of unrecognized truth, and I am sure that the same applies in the field of inquiry covered by this book.

ACKNOWLEDGMENTS

Apart from being a broad survey of the field, this book brings together much original work covering a period of some 30 years. I am very aware of the invaluable help given by so many people who helped to lay its foundations. I am particularly indebted to Dr. R. G. Scaife, who read the first manuscript and made many detailed comments and suggestions for bringing the manuscript to its present form. I would also like to thank those anonymous assessors who read the text and collectively made some most pertinent observations and helpful suggestions for improvement. In doing so they went far beyond what was incumbent upon them merely to discharge the task they had undertaken. I am most grateful to them.

1 SOIL POLLEN: PRINCIPLES AND TECHNIQUES

INTRODUCTION

The majority of archaeological sites are on land that is dry seasonally if not permanently; the types that are associated with waterlogged conditions are dealt with in Chapter 2. Pollen will only be preserved on land sites if some factor exists in the microenvironment that inhibits microbiological decay. As pointed out in the Preface, such a factor may be aridity, high acidity, or an excess of salt or toxic metal ions in the deposit. The commonest of these, at least in temperate regions, is high acidity (low pH), and as will be seen from later chapters this has made possible quite a substantial amount of pollen research on archaeological sites. In fact it is this factor above all others that has made possible the technique of soil pollen analysis (Dimbleby 1957, 1961a).

SOIL POLLEN ANALYSIS

On the Continent, soil pollen analysis has been used widely to correlate vegetation history with soil profiles. Without attempting to give an exhaustive list, and without detailing individual references, it may be noted that the technique has recently been used by the following: Bastin, Coûteaux, Dricot, and Munaut in Belgium; Géhu, Guillet, and Planchais in France; Trautman and Pertsch in Germany; Havinga in Holland; Circiumaru in Romania; Savina in Russia; Vuorela in Finland; and Welten in Switzerland.

In contrast, soil pollen analysis has been little used in England, apart from my own work, being distrusted by our conventional palynologists, mainly on the ground that soils lack perfect stratification. However, sporadic applications of the technique have been made by a number of people, mostly working in the context of moorland or upland soils.

As the interpretation of palynology of archaeological sites must depend on what happens to pollen in soil, this must be explained before we go further. It must be emphasized, however, that what follows can only be taken as applying to soils on moderate topography in a cool temperature region. Welten (1958) has shown that in the soils developing on steep topography in the Alps these processes are obliterated by other factors such as soil movement downslope. In arid areas processes again may be different, and Martin (e.g., 1963) has used pollen analysis of soils extensively in arid parts of North America where the deposits themselves may often be polliniferous despite their high pH.

Conventional pollen analysis makes use of a deposit that is progressively accumulating, such as a peat or lake mud, so that the pollen rain from the local and regional vegetation is incorporated as the deposit builds up. The assumption is therefore made that all the pollen at any given level is coeval. Though there are exceptions, it seems to be generally true that once incorporated within the deposit the pollen moves neither upwards nor downwards. If one samples serially through the deposit the fluctuations in the pollen spectra show changes that should represent a chronosequence. Such assumptions are today being questioned in certain circumstances: sloughing may take place in lake muds, or a basin site may receive pollen washed out of the soils of the catchment area, and it is known that errors can arise through contamination by pollen of different age becoming incorporated into the deposit. Nevertheless, in general acceptable chronosequences can be demonstrated.

INCORPORATION OF POLLEN IN SOIL

When pollen falls on a soil surface the course of events is normally quite different, to such an extent that pollen diagrams obtained from soils have to be interpreted in a totally different way. Except in regions with a long dry season, there is a predominant movement of water downwards through the soil and this brings about the downward movement of pollen. This downward drift of pollen, however, is not the simple movement of the pollen grains as discrete particles through the interstitial spaces of the soil. It can be shown that even in a soil very rich in pollen only a very small proportion is lying free in the soil (Dimbleby 1962b). This was once brought home to me forcibly in an investigation of the soil buried beneath the Roman Devil's Dyke near Newmarket, Suffolk. The archaeologist was anxious for a quick answer as to whether the soil contained pollen, so an extraction was rapidly made using caustic potash to destroy the humus. We found no pollen and reported accordingly. Some time later, in order to provide complete data for the final report, we reextracted the same samples, using acetolysis and hydrofluonic acid, and found enough pollen throughout the profile to give a pollen diagram (Fig. 1). It is clear that the pollen is locked up in the soil, perhaps in some form of humic complex, and is not free to move. Its ultimate release apparently depends upon the breakdown of whatever sort of aggregate is involved; movement is therefore extremely slow and is probably related to the rate of movement and breakdown of the colloidal humus component of the soil. The rate of downward movement no doubt varies with the pedological conditions, but some idea of its order can be obtained from the diagrams shown here. At Matley Wood (Fig. 2), for example, there is some indication that the present woodland was preceded by a pioneer phase of *Pinus* and *Betula*. The present tree cover was about 250 years old at the time of sampling, but the frequency maxima for pine and birch both fell somewhere between 1 and 3 inches (2.5 and 7.5 cm). Similarly at Pondhead (Fig. 5), where records tell us that the oak and chestnut were planted into grassland some 140 years before sampling, the grass pollen frequency also has a maximum between 1 and 3 inches of depth. On such evidence an estimate of perhaps 10 cm downward movement in 300 years would seem to be realistic. This is not to say that some grains do not move faster, as the curve for *Castanea* (sweet chestnut) at Pondhead shows (for further discussion see below).

It is important to grasp this point, for otherwise one would expect that small grains would pass through the soil more rapidly than large ones, so that any stratification would be merely a reflection of size. As it is, there is a crude stratification, with the oldest grains being the lowest in the profile, and certainly it is not true that these are always the smallest grains. I shall return to this crudeness of the stratification a little later, but to demonstrate it we must first look at other features of a typical soil pollen profile.

If the upper horizon of the soil is an A_0, an accumulating layer of organic matter derived directly from the vegetation, it can be treated in much the same way as a peat; that is, it is building up, so that at any one level the pollen is all coeval. As raw humus layers do tend to shrink and swell, there may be some movement of pollen down cracks, but acceptable analyses have been carried out on deep layers of this type (e.g., Iversen 1964). In the underlying mineral soil, however the pattern changes, and this is reflected not only in the percentages of the different pollen types but also in the absolute pollen frequency of the different taxa in each sample, that is, the actual amount of pollen of each type in each sample. Confusingly, the *percentages* (based on various denominators) are often referred to as *frequencies*, and it is important to keep the distinction clear. From my own analyses of soil pollen it is always possible to calculate both absolute frequencies and percentages, and a glance at Figure 2 will show that they are very different parameters.

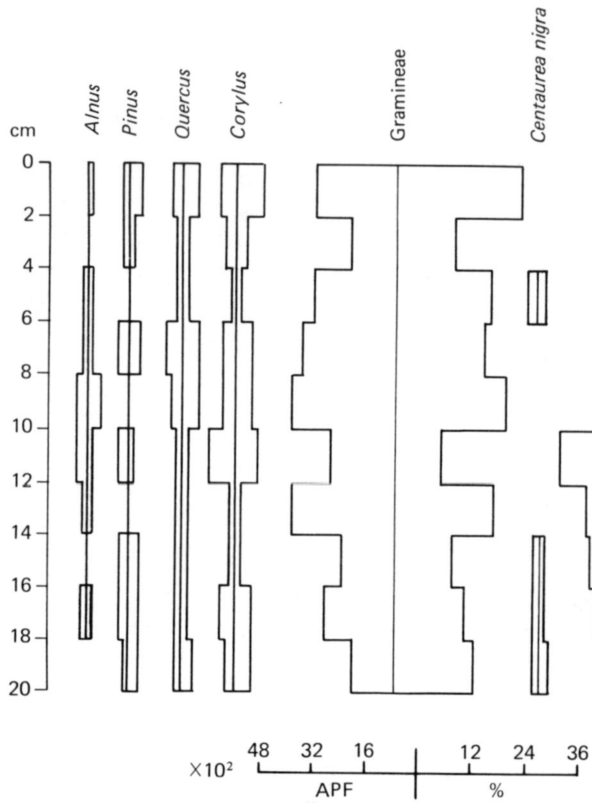

Figure 1. Pollen diagram from a soil profile beneath Roman Devil's Dyke, Newmarket, Suffolk. APF, absolute pollen freguency.

POLLEN DISTRIBUTION IN THE PROFILE

In an acid soil such as a podzol or *sol lessivé,* in which soil-mixing organisms such as earthworms are not present, there is a characteristic distribution of pollen with depth. The absolute frequencies are high near the surface, on which the pollen rain falls, and decrease with depth. This feature is of great importance in the recognition of buried surfaces (see Ch. 3); however, the quantities of the different taxa are not solely related to depth but also to age. The longer pollen lies in the soil the more it is attacked by bacteria, particularly actinomycetes, so the actual number of grains declines, but also the greater is the depth to which the pollen is washed. Consequently there is

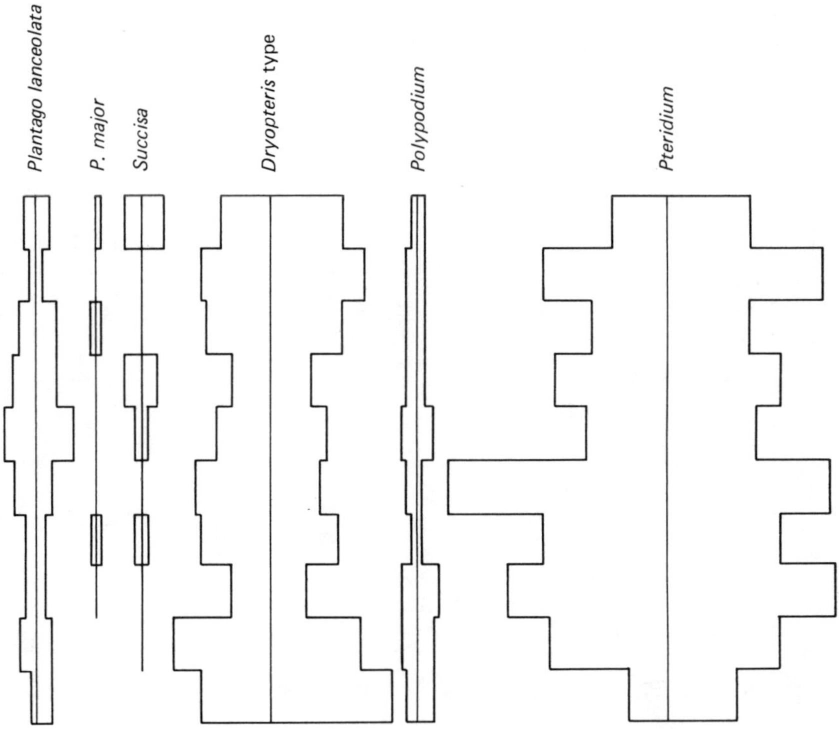

Figure 1. *continued*

normally an overall decline in the total amount of pollen with depth, but for the older taxa a less marked decline or even an increase in actual frequency in the lower layers. Percentage curves can give a distorted impression of the abundance of the older pollen types. Figure 3 illustrates a hypothetical situation covering three taxa: one recent, one ancient, and one of intermediate age. The actual frequencies are shown in the curves on the left, but when calculated as percentages the curves on the right of the axes are obtained. Unless frequencies are represented, the percentage curves alone may be grossly misleading. Notice that the peaks of the intermediate pollen curves do not occur at the same depths.

These curves also demonstrate another feature that is a major point of difference in comparing soil pollen diagrams with those from stratified deposits. Although the bulk of the recent pollen is in the top 4 cm of the surface of the soil, some grains have percolated well down into the profile. At the same time, the bulk of the ancient pollen is in the lower layers of the profile, but there still remains a remnant in the upper levels. The pollen of intermediate

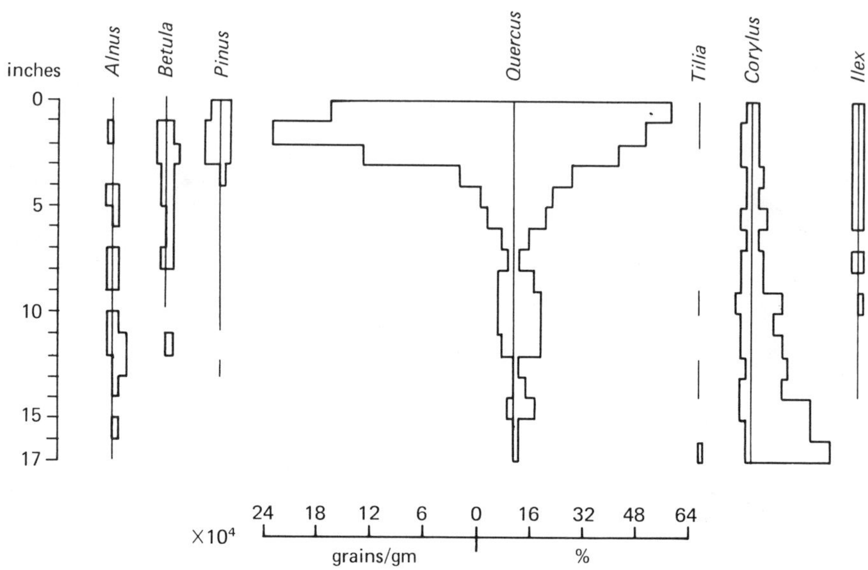

Figure 2. Soil pollen profile from Matley Wood, New Forest, Hampshire.

age is spread over a wide range of depths, though its peak abundance is between the other two.

A soil pollen profile is not therefore truly stratified. At any one depth there is pollen of various ages, which would not be the case with stratified deposits. The pollen diagrams from soils therefore have to be read in quite a different manner; this has not been understood by those who have criticized soil pollen analysis on the basis of its lack of precise stratification. Many years ago I found a Late-glacial element in the pollen at the base of a moorland podzol; knowing Professor Godwin's interest in this period I sent him the full analysis. To my surprise he used the data as a basis of a paper about the principles of soil pollen analysis (Godwin 1958), in which he recognized the existence of downwash. However, he found it necessary to interpret this as an episodic process, with intervening periods when pollen was being rapidly decomposed. Apparently he did not see downwash as an ongoing process, though I see no difficulty in this and no necessity for the complicated pattern of processes he suggested. He did not recognize the existence of a buried soil within the podzol profile (Figs. 21 and 22; see also Ch. 5), which is of fundamental importance to any interpretation.

Clearly the forms of the curves can tell us something, at least in broad terms, about vegetation succession at the site. At the same time Figure 3 makes it obvious why a spot sample from a soil can be of only limited value or

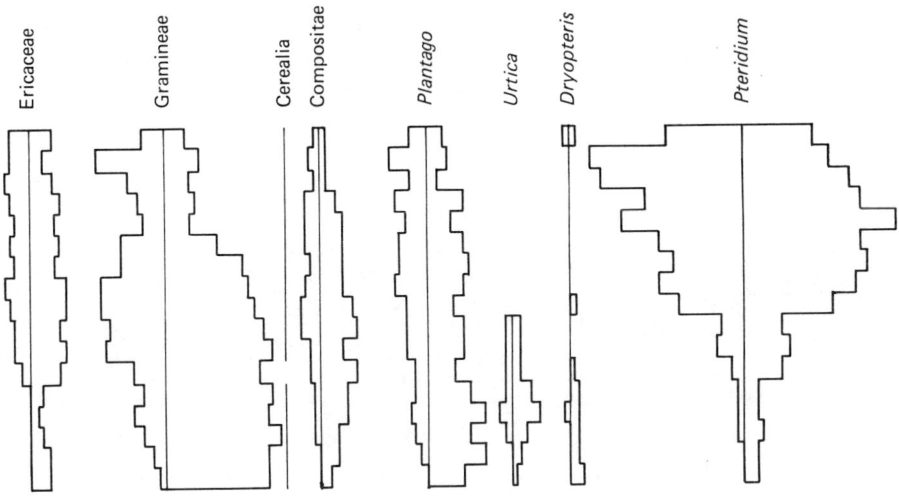

Figure 2. *continued*

even misleading. Such samples are only trustworthy when they come from near the surface of the profile where the abundance of contemporary pollen effectively swamps that of earlier stages. Some have argued that at low frequencies the curves are not reliable. This is based on a failure to recognize that though the absolute frequencies may be very low in the lower layers of some profiles, counting has to be continued until adequate totals have been

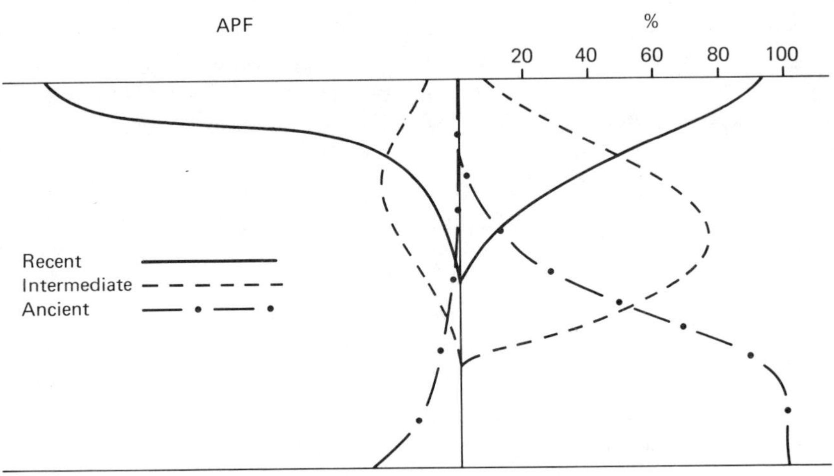

Figure 3. Curves showing the theoretical distribution in soil of pollens of different ages.

obtained, just as would be done in conventional pollen analysis. In my work on soils a total of at least 200 is arrived at. This may mean searching a larger number of traverses on the microscope to get the requisite total, and if this proves impossible the total on which the calculations are based should be specified. Generally speaking where adequate counts cannot be obtained the analysis should be stopped, despite pleadings from a hopeful archaeologist.

There is another reason for discounting analyses where there is difficulty in obtaining totals of at least 200, because of low absolute frequencies. The low frequencies as we have seen, usually occur in the lower levels of the soil where what pollen is present is residual and often of great age—possibly thousands of years old. It is what is left over after the pollen grains of an earlier period have been subjected to microbiological attack for a long period of time. Different pollen types have different degrees of resistance to decomposition, so that the types which survive are those most resistant to decay. A glance at any soil pollen analysis will show that the upper layers of the soil have many more taxa than the lower ones; many of them decay before reaching the lowest levels. Consequently onc finds high percentages of fern spores and other resistant grains at the greater depths; clearly they cannot be interpreted directly as an assemblage indicating a particular plant community. But high percentages should be checked against the absolute frequencies at these levels. Differential decomposition cannot *increase* the amount of a resistant type in the soil (though frequencies can increase if the pollen is piling up against a barrier such as an iron pan), yet one frequently finds that there is an actual increase in the lower layers even of a free-draining soil. Where this is the case it is reasonable to see the 'ancient' component as the remains of a pollen rain not represented in the upper parts of the profile. Over the years I have found many different taxa dominating the lower layers of soil profiles; not only fern spores, but grasses, heather, hazel, alder, oak, and lime (Dimbleby 1962b). Some of these are clearly indicators of well-defined vegetation types, so that where they occur their presence is some indication of what had previously occupied the land (e.g., Fig. 2).

The question of differential decomposition has been investigated experimentally by Havinga (1964, 1967, 1971). He buried the pollen of some critical taxa in five different soil environments and studied the changes over a period of several years. It was apparent that there was a big difference in the conditions for preservation between the soils, a river clay being much less favourable than a podzolized sand. There were also marked differences between the different taxa; spores of *Lycopodium* and *Polypodium* were generally the most resistant to decay, but the pollens of the commoner tree species varied relatively among themselves. A surprising fact that emerged was that the order of resistance was not always what one would expect from field evidence. Hazel and alder pollen, for instance, were found to disappear quite rapidly. This is not in accordance with field experience, as can be seen from a number of the diagrams published here.

Oak in particular raises a problem, not only in these experiments but also in field analyses (Havinga 1964, 1974). Oak pollen seems to be unaccountably absent from some deposits where one would expect it to be present, for instance where previous woodland is indicated by the presence of pollen of other tree or shrub species. In other cases oak pollen can be abundant. Figure 4 shows a previously unpublished analysis of an old oakwood site in Sutton Park in the English Midland (for soil details see Mackney 1961), which contrasts with the profiles from the North York Moors (Dimbleby 1962a). Other authors have also observed differences from one area to another, in materials which show no obvious fundamental differences. It is possible that the question is one of ecology and history. Oak (most pollen analysts have difficulty in separating pedunculate and sessile oak on their pollen) is a tree with a wide range of soil tolerance, from strongly calcareous soils to moderately acid ones. If we are examining soils that have undergone a major transition from, say, woodland to heath, the initial starting point may be the clue. Some soils carrying heath today were originally sufficiently base-rich to have carried a mixed oak forest; others were less so, carrying only an acid oak forest. In the former case the bulk of the oak pollen over the ages would have fallen on a soil of relatively high pH and been destroyed, the only tree pollen surviving being that of species that formed the transition from woodland to heath as the acidity increased. On the soil which had always had a low base exchange capacity the oak pollen would have been preserved even during the forest phase and would therefore have persisted in large quantity in the heathland soil. Certainly this would explain the cases quoted above.

It is apparent that there is no rule of thumb that can be applied to all cases. The interpretation of soil pollen requires a knowledge of ecology in the field, and is a hazardous business for those without that background.

SOIL HETEROGENEITY

Other criticisms sometimes levelled at soil pollen analysis are based on a belief that soil is so heterogeneous and and so biologically active that there can be no meaningful pattern in the soil pollen.

The presence of persistent channels of drainage in a soil is sometimes cited as a factor which will render a pollen profile invalid. It is, of course, true that there are such channels, not least those left behind by decayed roots. It was explained above, however, that the pollen in soils is mostly locked up in some form of humic aggregate and is not free moving. Nevertheless, a small fraction of recent pollen may be moved downwards before it becomes locked up,

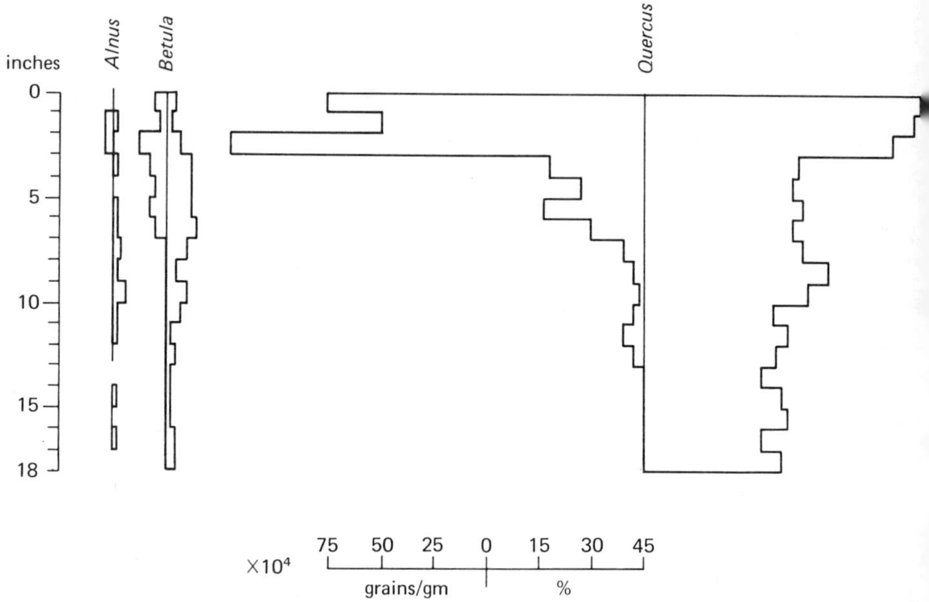

Figure 4. Pollen profile through an ancient oakwood soil, Sutton Park, Warwickshire.

and it may do so down channels of free drainage. When this occurs, however, it is localized to a channel, and whilst it may be impossible to avoid such channels in sampling, when they are analysed they can clearly be seen to have an extraneous content of later pollen which contrasts with the main mass of pollen in the sample. A good example is seen in the mesolithic levels at Oakhanger (Fig. 28), where one sample contains grass and cereal pollen, together with an increased amount of *Calluna* (heather), clearly derived from later levels higher up, but it presents no problem to the interpretation of the profile as a whole (Rankine *et al.* 1960).

At Pondhead Inclosure, in the New Forest, similar contamination was found (Dimbleby 1962a), though this was not associated with an archaeological site. The pollen profile developed in an acid sand clearly shows two phases, one superimposed upon the other (Fig. 5). Such a phenomenon is not unusual and is discussed more fully in Chapters 3 and 5. The upper sequence shows forest clearance, through agriculture, to heathland, whilst the lower one represents closed deciduous forest. It seems that a forest soil has become covered with blown sand, causing a new pollen record to develop from the new surface. The point of interest here is that though the buried forest soil contained the merest traces of grass and weed pollen, cereal pollen grains

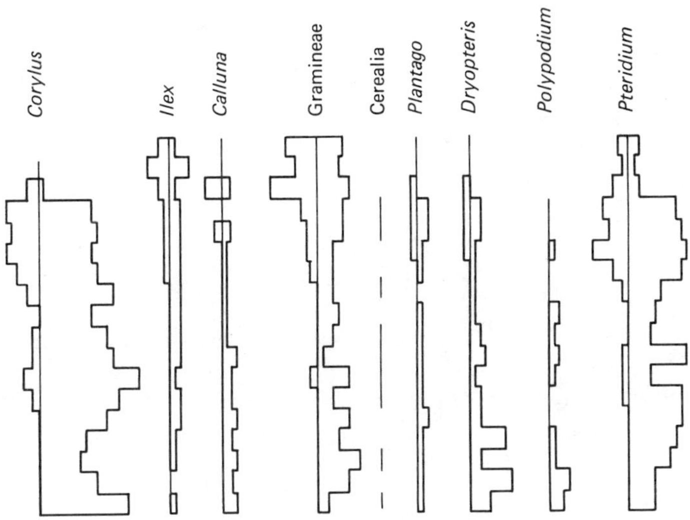

Figure 4. *continued*

were recognized from four of the samples from this layer. They are clearly intrusive and probably reached this lower level *via* root channels. Perhaps most conclusive was the trace of *Castanea* pollen in this buried soil. As *Castanea* was not planted until A.D. 1810, this pollen could only have come from the present surface. But this contamination was easily recognized for what it was and in no way affected the interpretation of the profile as a whole.

FAUNAL MIXING

Activity by soil fauna is very different in its manifestations, and can even provide positive environmental evidence about soil history. Vertebrates burrowing in soil of course create havoc but as their disturbance is so obvious there is no problem in avoiding it in sampling. The effects of soil-mixing invertebrates are more difficult to avoid, but they present much less of a problem than might be thought. Those which cause the most disruption are the fauna associated with a mull type of humus, and on the whole such soils tend to be circumneutral and not acid enough to preserve pollen well. It is the more acid *sols lessivés* and podzols which are most commonly used for pollen analysis and in these the soil mixing fauna as a group are insignificant. Some species of earthworms, millipedes and other humus-feeding arthropods may be present, but they mostly work within the surface humus layer and do not mix the humus with the mineral soil beneath. The species which do mix,

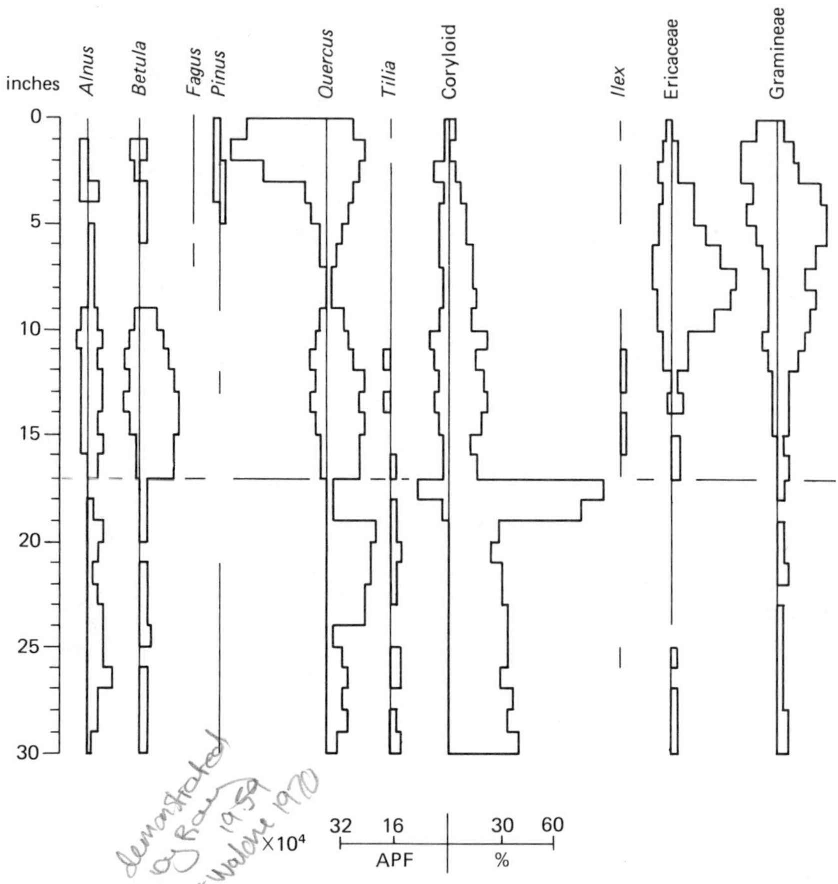

Figure 5. Pollen profile through a podzol at Pondhead Inclosure, New Forest, Hampshire.

notably certain species of earthworms and some millipedes, are inhibited by pH values below about 4.5. However, pollen can be found in some soils with pH values of 4.5 or even higher, and when that happens the fauna effectively destroy any pattern of pollen distribution. Ray (1959) showed in a simple pot experiment that large earthworms, supplied with pollen at a soil surface, will quickly distribute it throughout the humus layer in which they feed and cast. In a more sophisticated experiment Walch *et al.* (1970) showed that pollen spread in layers under a soil cover was carried both upwards and downwards when earthworms were present, but only moved downwards in their absence. This downward movement was attributed to rain percolation. They also dem-

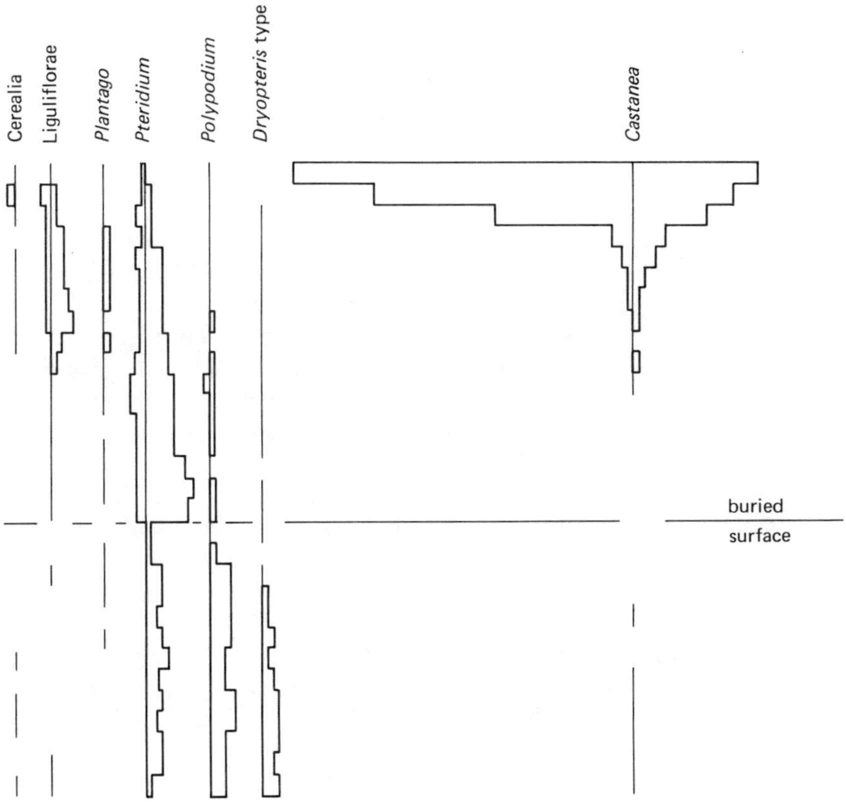

Figure 5. *continued*

onstrated that radioactively marked pollen from a buried layer could eventually reach the atmosphere, *via* surface worm casts which dried out and were then picked up as dust by the wind.

Clear evidence of the movement of pollen by soil fauna was also obtained from two experimental earthworks—each a linear bank and ditch—built on a rendzina (Overton Down) and a heath podzol (Morden Bog) respectively (see also Ch. 3). Before each bank was built the soil surface was dusted with *Lycopodium* spores, alien to the local flora and very resistant to decay. Over the course of a few years the spores had migrated both downwards and upwards in the calcareous milieu, but remained at the level of the buried surface in the podzol (Jewell and Dimbleby 1966; Evans and Limbrey 1974; see also Fig. 14). In the long term earthworm action mixes the pollen throughout the humus zone within which the worms are feeding. This is easily recognized in pollen profiles by the fact that both the frequencies and percentages of the various pollen taxa scarcely change with depth (e.g., Dimbleby 1962a:82; Dimbleby and Evans 1974).

Even when the pollen in soils has been mixed by earthworms or other agencies useful information may be obtained. Working on Dutch soils developed in cover sand Havinga (1963, 1974) has distinguished between different types of pollen profiles. One is what he calls a homogenized profile, in which complete mixing has been brought about throughout the profile by the soil fauna. If there is a change of vegetation without a change in the depth of biological activity the new pollen types will be distributed throughout the profile, along with, or in place of the original pollen assemblage. If on the other hand there is a reduction in the depth of incorporation, then the profile will show two zones, an upper one containing the new pollen, and a lower one containing only the older assemblage (Fig. 6). Havinga also recognizes the pollen profile that is dependent purely on the downward movement of pollen under the influence of percolating water (whatever the micromechanism of this process), that is, profiles of the type described above. When there has been a change of soil conditions, homogenization may be replaced by downwash, as for instance when a deciduous forest is replaced by heath. In such a case a brown-earth soil with earthworms may be replaced by a podzol too acid for the survival of the worms.

There are various factors that can bring about the sort of changes described by Havinga. One which he mentions is change in the level of ground water, a factor to which much land in low-lying Holland is subject. However, major pedological and ecological changes can take place under many site conditions,

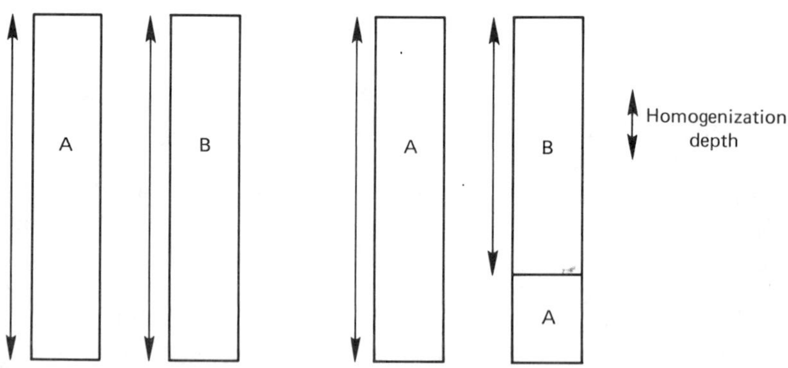

Figure 6. Change of a pollen profile under the influence of homogenization: A, pollen composition representative of period A; B, pollen composition representative of period B. On left: the homogenization depth is the same during period B as during the preceding period A. On right: the homogenization depth is less during period B than during period A. (Reproduced from Havinga 1973.)

particularly on base-poor soils which, as we have seen, lend themselves to soil pollen analysis. In fact it was this field of research that first led me to soil pollen analysis. Pollen data from archaeological sites may shed light on the question of soil change, and as this problem will emerge again in Chapter 3 it is appropriate to give an outline here. The distinction between what Havinga calls homogenized soils and soils in which downwash is the main cause of pollen movement is also the key to interpreting soil pollen in terms of soil processes.

Soils undergo development (maturation) with time and, ignoring any climatic change during the period in question, this means that in the moist cool temperate climate of Atlantic Europe there is a continuous trend to greater acidity by the removal of soluble bases by percolating water. Some of these bases are vital plant nutrients, or are important in maintaining the base status of the soil, which in turn determines the organisms which can work in the soil and on which the soil processes ultimately depend. The parent materials from which soils develop vary greatly in their reserve of such bases. Chalk, for instance, is almost entirely composed of calcium carbonate, and though there may be massive removal of calcium by solution (leaching), there is still an inexhaustible supply of this element in the weathering parent material. Consequently, under the influence of percolating water alone, soils on chalk change very little with time, and remain 'mixed' soils.

Such is not the case for soils developed on parent materials with low base status. Here, soluble bases are not, or only slowly, replaced, so the soils become more acid. At the end of the last glaciation some parent materials were moderately well endowed with base reserves; some soils of the earlier part of the postglacial period were therefore relatively rich biologically. They had a circumneutral pH, and could harbour the soil-mixing agents such as earthworms. But as the postglacial proceeded they became progressively more acid as the bases were leached out, the mixing fauna disappeared and the soils ultimately became podzolized. This process may account for the pollen pattern in the buried soil at Cock Heads (Fig. 21), where the Late-glacial pollen is distributed throughout the depth of the buried soil, whereas the later pollen is not uniformly mixed. It has been shown that similar trends in acidity occurred in earlier interglacials (before man exerted a recognizable influence).

Ball (1975) has discussed the development of soils on low-base parent materials in the Highland Zone of Britain, paying particular attention to the influence of man. He concludes that man had the effect of altering the rate of pedogenesis, the progressive acidification described above, without altering the actual sequence of stages. He sees no evidence that man created any new types of soil. The study of soil profiles and pollen sequences beneath earthworks suggests that the clearance of the forest (often with a 'homogenized' brown forest soil) accelerated the process of acidification (Dimbleby 1962a, 1978c); it can be shown that differentiation of podzol profiles took place within a very short period (in pedological terms), perhaps a century or two. The earlier soils were often poor or deficient in pollen, the later ones relatively pollen-rich, doubtless due to the significant drop in pH. Romans and Robertson (1975) have used micromorphology and charcoal analysis on buried prehistoric soils and have been able to trace a general trend from brown forest soils with oak predominant to more podzolized soils associated with slash-and-burn clearance and subsequent land use.

In my opinion the accumulating mass of evidence from prehistoric soils does not substantiate the conclusions reached by Ball. It is agreed that forest clearance and the use of land for primitive agriculture did accelerate soil acidification, but it is significant that over a wide area of our upland soils the soil type which developed was the stagnopodzol, very acid with a shallow but indurated iron pan. Such a soil is incompatible with forest conditions, and I cannot see it as part of a chronosequence of progressive soil acidification through the postglacial period. Unfortunately, Ball misrepresents my belief, for he thinks I am saying that trees could not survive on the soil types beneath barrows. I see the lack of trees as due to land use, particularly grazing and burning. Indeed, even the most degraded of our modern soils seem able to support a tree succession if these factors are halted. We have pollen evidence from buried soils suggesting that woodland regeneration sometimes took place.

Even today colonization by trees can take place and halt or even reverse the continued acidification—at least in the short term (Dimbleby 1952). This happens, for instance, when heathland becomes colonized by deciduous trees for long enough for soil changes to become established, that is, a process of regeneration of both vegetation and soil. Under these conditions the upper horizons of a podzol, in which the characteristic distribution pattern of pollen downwash occurs, will be worked by newly invading soil fauna, so destroying that pattern. Again, this fact itself is a valuable pointer to what has been taking place, regardless of the pollen spectra themselves.

Such changes in the biological activity of soils may be brought about by artificial as well as natural agencies. For instance, the addition of lime-rich material (road metalling, collapsed plaster work, fertilizers, etc.) on the top of a previously acid soil may initiate just such changes. I have seen such a change beatifully illustrated where limestone metalling or a collapsed wall has spread limestone over the surface of a moorland road (Dimbleby 1952). Not only will this lead to mixing as soon as earthworms appear, but if the pH is maintained high enough for long enough the resultant microbiological activity may lead to the destruction of both old and new pollen throughout the zone of activity.

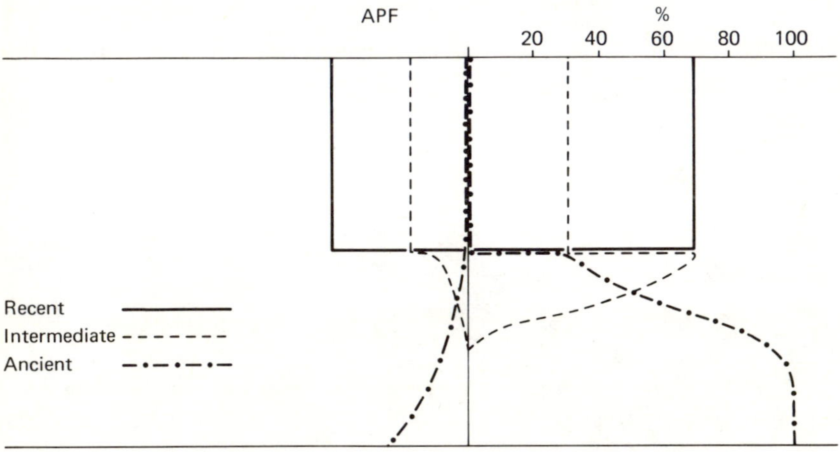

Figure 7. Effect of homogenization on theoretical soil pollen curves (see Fig. 3).

As an indication of how the mixing of the upper layers of a soil will affect the pollen distribution, take the set of curves shown in Figure 3. On the assumption that the top part of the profile is completely mixed (and for the sake of simplicity leaving out any newly introduced pollen) the resulting distribution will then be as shown in Figure 7. The figure shows the uniform distribution in the upper horizons contrasting sharply with the unchanged distribution below. We shall be talking in detail later (Chapters 3 and 5) about disjunctions in pollen profiles. Commonly they are associated with buried surfaces, but in this case a disjunction would have been produced within a soil by natural processes. However, mixing can also be brought about by artificial means such as ploughing, which, if repeated regularly can have an effect on the pollen pattern difficult to distinguish from the mixing by soil organisms.

ARTIFICIAL CONCENTRATION OF POLLEN

Sometimes one finds unaccountably high concentrations of a particular pollen type—far more than can be accounted for by the normal processes of soil development. Examples of this will be discussed in Chapters 5, 6, and 7; at this point it is only necessary to mention that pollen can get into soils through agencies other than those already mentioned, for instance, man may bring flowering vegetation onto a site as part of his agricultural or husbandry practice and this will then lead to an exceptional accumulation of the pollen of that vegetation on that site. Apart from human agencies some animals can lead to concentrations of pollen at certain points in the soil; for example ground-nesting bees may collect pollen and bring it to their subterranean nests (again see Ch. 7). It is possible that some foraging ants may also produce a similar effect.

SOIL MOVEMENT

An important element in the study of archaeological sites and their deposits is the mass movement of soil. This may be brought about by erosive agencies such as wind or water, or by man's own action in cultivating the soil; frequently these are closely connected. We have seen that the pollen is locked up in the soil microstructure, so that when soil is moved *en masse* its pollen moves with it. This is of great importance in studying buried soils (Ch. 3), for there may be a contrast between the pollen in a buried land surface and the overburden which may give a clue as to the causes of burial of the old land surface. Nor must it be forgotten that man may deliberately shift soil, as in the building of earthworks. Here too the pollen content of different strata can be very informative (see Ch. 4).

POLLEN-POOR SOILS

So far we have been discussing soils in which pollen is present in reasonable quantity, at least in the upper layers. These are generally acid soils. However, many archaeological sites occur on neutral or base-rich soils. In fact such soils would be of more value to man than acid soils for they are more productive. It should not be forgotten, however, that as pointed out above many of the soils we now classify as acid were originally more base-rich and have become impoverished through man's (mis-)use of them (Dimbleby 1962a). What are the possibilities of applying pollen analysis to base-rich soils, such as the calcareous soils derived from chalk or limestone? In the early days of this work I thought this a hopeless proposition, for such soils are biologically active and consequently the pollen is destroyed. It came as no surprise, therefore, to find that many such soils are quite devoid of pollen. However, as such soils are of great archaeological interest, I found myself subject to much pressure to examine buried land surfaces for pollen; in a few cases I rather disconcertingly did find some. Then the pressure from archaeologists increased.

Where pollen does occur, it is always in very low concentrations—at most a few thousand grains in a gram of soil. One practical effect of this is that the pollen counting becomes very laborious. It may be impossible to concentrate the pollen by standard procedures because of insoluble particles of clay or silt making it necessary to traverse 20 or 30 slides for each sample in order to reach a statistically valid total. Even this often cannot be achieved, a fact that the disappointed excavator sometimes has difficulty in accepting. Assuming one can reach an adequate pollen total (200) for each sample, how meaningful are the results? Generally, of course, there is very little stratification of the pollen due, clearly, to the mixing of the soil by the fauna, particularly the earthworms. It has been noticed however, that although microbiological decomposition is very active in such soils, the condition of what pollen there is may be very good; the grains do not appear to be in an advanced state of decay. A possible explanation of this may be that all the pollen is very recent-perhaps less than a year old—and has not yet succumbed to microbiological attack (Dimbleby and Evans 1974). If so, this could be very useful, for the grains could be taken as indicating the pollen rain over a very precise period. Moreover, as the soil is well mixed it should be allowable to work on the pollen sum for the whole profile, thus rendering it unnecessary to reach an adequate count in each sample.

In calcareous soils, however, there is another means of investigating the ecological history. Small land snails are present which are good ecological indicators and as their shells remain for a long period in the soil and can show some degree of stratification, they can give similar information to pollen. Sometimes the evidence from the molluscs appears to be at variance with that from the pollen, in which case they are probably derived from different phases of the ecological history. For a fuller account see Dimbleby and Evans (1974).

By the use of special techniques it is possible to extract pollen from materials with only a few hundred grains per gm (Frenzel 1964). This has been done, for example, to extract pollen from the Hoxnian interglacial deposits at Swanscombe, Kent (Hubbard 1982). Using a variant of Frenzel's method, Hubbard was able to obtain a satisfactory analysis of deposits with a pollen frequency as low as one grain per gm. It is necessary to establish a probability that the pollen in such low concentrations is in fact derived from contemporary pollen sources. The possibility that secondary pollen is present, perhaps residual from a different geological era, always has to be borne in mind, for such pollen will be concentrated by these techniques just as will the contemporareons pollen.

However, Hubbard argues that these low concentrations are more likely to be a representation of the contemporary pollen rain, on the grounds that in such actively corroding soils more ancient pollen and spores will disappear first, leaving only contemporary pollen. Personally I have reservations about this argument; though I have not made relevant observations on calcareous soils, I have noticed that in more acidic ones reworked palynomorphs are recognizable by being darker in colour. If this darkening is due to an age-related change such as polymerization of the material of the pollen grain wall, this may alter significantly the susceptibility of the material to decomposition in the soil, so altering the basis for our usual assumptions about relative rates of breakdown.

OTHER POLLINIFEROUS DEPOSITS

The preservation of pollen in calcareous soils as described above is characteristic of moist temperate climates which favour microbiological activity in the soil, but as was pointed out at the beginning of this chapter, there are other factors which can come into play. In arid regions, for instance, microbiological activity may be arrested by the lack of water, and under such circumstances pollen can be preserved even in calcareous soils (e.g., Martin 1963).

Other terrestrial deposits which have been subjected to pollen analysis occur in caves or rock shelters, giving special problems of interpretation. These will be dealt with in more detail in Chapter 6. The palynology of waterlogged deposits is standard practice and will not be described here. Details may be found in textbooks on palynology (e.g., Birks and Birks 1980; Moore and Webb 1978). Archaeological sites associated with waterlogged deposits are more easily studied palynologically than terrestrial sites, though they are not without problems of interpretation. They will be considered specifically in the next chapter.

SAMPLING

Having explored the occurrence of pollen in deposits of archaeological significance, the next question is how to extract and analyse it. The point was made above that because such deposits are generally not stratified in the precise sense of the term, any one sample is likely to contain pollen of various ages. As a general rule, therefore, spot sampling is valueless. The only situation in which it is acceptable is where there is no doubt that the pollen in the sample can only have originated at one time; an old land surface, or a buried turf, are examples of such a situation.

In any other context sampling has to be serial; that is, a sequence of samples is taken through the deposit, from which pollen curves can be built up to provide the basis for interpretation. In deep deposits the samples may be spaced at intervals of approximately 5 or 10 cm, but where an archaeological feature appears it is usually necessary to sample more intensively, perhaps as closely as 0.5 cm, below and above that feature. This allows the recognition of short-term environmental change, perhaps on a localized scale, associated with that feature.

In normal soil profiles contiguous sampling is necessary; that is, the samples should touch each other so that a column through the soil profile is completely sampled. One reason for this is that soils are often compound, with an old soil becoming covered over so that the pollen rain then descends onto a new surface. Contiguous sampling makes it possible to locate the original surface with precision (see Chapters 3 and 5).

Some of the early work on soil pollen, including my own, was based on sampling by horizons. This, however, can and did produce nonsense results, for in many cases the horizons we see today are of relatively recent development and may in fact postdate much of the pollen in the profile (Dimbleby 1957, 1961a). As soon as contiguous sampling was applied, however, sensible sequences emerged.

It is not possible to give instruction on sample size and technique which can be regarded as standard in all cases. We have seen that the abundance of pollen in a sample will vary greatly, but of even greater practical significance is the physical nature of the deposit itself. A technique which works well on a soft peaty material may be quite useless on a gravel. In soils the texture can change dramatically from the top to the bottom of a profile.

In those materials suitable for pollen analysis, the pollen grains occur in very considerable numbers, often reaching hundreds of thousands of grains in a gram of soil. Since the number of grains required for a statistically acceptable count varies between 200 and 1000 (see below), it is apparent that only small samples are necessary. It has been my custom to take samples equivalent to a matchboxful (ca. 30 gm). Sometimes archaeologists submit samples of several hundred grams in weight but these are almost always useless, for they are clearly 'bulk' samples and not taken from narrowly defined levels in the section. As such they could cover several distinct pollen 'episodes', as reference to Figure 2 will indicate.

It is a prerequisite of all sampling that it should be done from a freshly exposed face, to avoid contamination by airborne pollen or secondary pollen contained in blown dust. Pollen can also be carried by water running down a face, so it is desirable to trim a face back a good 5 cm before sampling and clean it by sideways scraping. If the deposit to be sampled is organic, or is a quite stone-free silt or sand, it can either be cut out in small blocks or sampled with a tubular borer. This can be a simple glass tube, open at each end, which is pressed in to the deposit, extracted and then corked up at each end. Such a borer is ideal for interval sampling, but for contiguous sampling the cut block method is preferred. Where the material to be sampled has a high clay context the cut block is usually the only possible method. Gravels and stony soils, on the other hand, do not lend themselves to this technique, and these comprise a considerable proportion of archaeological site materials. The stone fraction itself is only of interest if one is concerned with the actual pollen frequency of a given volume of deposit. For instance, many soil profiles become increasingly stony as depth increases. It has been pointed out by Rennie (1958) that pollen frequencies are being overestimated if one bases calculations only in the fine fraction, neglecting the stone fraction. Another way of looking at this is that as one goes down such a profile the proportion of fine material in the soil commonly falls. For practical purposes this consideration can be ignored where it occurs, for even on the basis of the fine fraction alone the frequencies, as we have seen usually fall steeply with depth and the effect of taking the stone fraction into account would only be to make this more pronounced.

In this context the absolute frequency *per se* has no particular value—many variables are at work—but the trends are significant. Rennie makes the point that for pollen-rich soils corrections for stoniness are probably unnecessary, but that at very low pollen concentrations there may be a significant effect. However, as we have seen, low concentrations are often associated with high microbiological activity and the presence of earthworms. In such soils absolute frequencies have very little meaning in themselves, except in so far as they demonstrate soil mixing. Rennie does suggest, however, that by taking stoniness and density into account features such as buried surfaces could be shown up with much greater clarity. I myself have not pursued this suggestion, but it is well worth further exploration.

In stony soils or deposits, we therefore have the problem of separating the fine material from the stones. As the stoniness can make it extremely difficult to remove a sample of requisite size, the separation is best done at the time of sampling. An effective tool for this purpose can be made out of a piece of copper (malleable yet hard enough for probing stony soils) shaped as shown in Figure 8 and folded inwards along the lines indicated. This gives a tube of rectangular section, with a cutting point at one end; by fitting a plastic bag over the other end the fine material dislodged by the point falls down into the bag. An effective method with this tool is to score on a cleaned face the chosen line of sampling, alongside which a scale is placed, and then to work up the face at the required intervals, removing the fine materials from amongst the stones. Using this technique—or indeed any other method of contiguous sampling along a selected vertical line—one should always work from the bottom upwards. This avoids contamination of the unsampled part of the profile by falling debris, which is particularly important in soil profiles in which the upper layers are the most pollen-rich. A profile after sampling with this tool is shown in Figure 9.

It should be unnecessary to emphasize to archaeologists the importance of labelling and recording during sampling. Each sample must be clearly labelled in such a way that the label will still be readable by the time the sample is processed. As this time should be as short as possible to avoid micro-

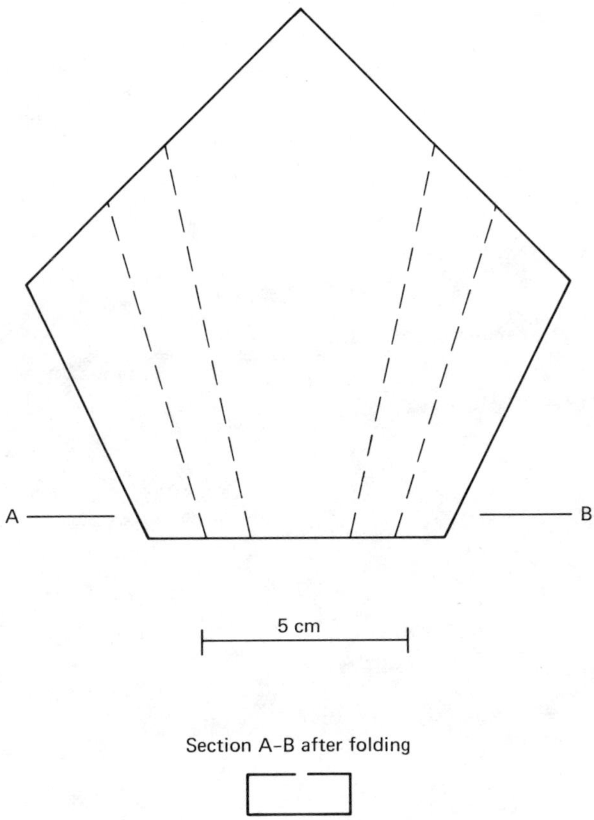

Figure 8. Template for the construction of a soil sampler.

biological damage to the pollen, printed paper labels (e.g., cloakroom tickets) put in the sample bag often suffice, but if samples are very wet, or highly humic, such labels can soon become damaged or illegible, in which case external tags should be used. Cross-referencing to a profile description is essential, and this description should include any visible features which might be significant in relation to pollen distribution or preservation. Examples of such features may be humus-rich layers, iron pans, root channels, the presence of small land-snails (indicating high pH), evidence of earthworm activity, layers of artifacts, charcoal, and so forth.

Once samples are taken they are susceptible to microbiological attack. Ideally they should be dealt with in the laboratory within 48 hours, but this is not always possible. Chemical methods of preventing decay are not very reliable. The use of a vaporizing disinfectant such as toluol may be effective

Figure 9. A soil section after sampling with the sampling tool.

on dryish samples, but in moist ones does not permeate all the soil aggregates, and it is the moist ones that are at most risk. Though storage at 4–5°C or even deep freezing can be useful, the most effective treatment is drying. Some pollen analysts, however, believe that drying can adversely effect the pollen by causing distortion. I am personally not convinced of this, and in any case pollen analysts dealing with mineral deposits are accustomed to dealing with badly deformed or broken pollen. Pollen in pcat may be in a beautiful state of preservation (though not always) but in soils and mineral deposits it is subject to many soil processes, including drying/wetting, freezing/thawing, and the operations of soil organisms. It is remarkable that pollen can be ingested by earthworms and even larger animals and be excreted in a recognizable form, though somewhat battered in the process. Even pollen-feeding animals such as bees excrete the pollen exines without destruction of the features necessary for recognition. The cell contents, of course, will be digested, but the indigestible outer coats, on which recognition depends, are excreted.

To prevent fungal and bacterial growth on the samples it is necessary to air-dry or oven-dry the samples. In humid atmospheres air-drying may not be sufficient. The safest practice is to oven-dry the samples at 90–100°C, hot enough to dispel the moisture, but not hot enough to damage the pollen. Pollen walls blister and distort if heated to too high a temperature. If oven-drying is not practicable (makeshift can be made with a domestic oven, leaving the door ajar), sun-drying may be used, though in this case strict precautions must be taken to avoid contamination by airborne pollen and dust.

PROCESSING

Once the samples have been taken, adequately recorded and dried, the archaeologist has done as much as he can. From then on pollen analysis becomes a job for the specialist, requiring the facilities of a laboratory. If, as is usually the case with mineral (as opposed to organic) samples, hydrofluoric acid has to be used, not only is special equipment necessary, such as polythene fume-cupboard windows and alkathene waste pipes, but extreme care has to be taken by the operator handling this most dangerous chemical, which can damage skin, eyes and lungs.

Details of a processing procedure which has been developed over the years are given as an Appendix. I do not propose to amplify them here. It has been suggested to me that by not doing so I am short-changing readers who might legitimately expect to find full details in a book of this sort, but I am unrepentant. This book is directed primarily at the archaeologist, who would normally expect such specialist work to be done for him; those with adequate scientific and laboratory experience will have no difficulty in finding appropriate details in other works (e.g., Moore and Webb 1978; Faegri and Iversen 1975; Birks and Birks 1980). There is a danger in the unqualified launching into a field like this and I do not want to encourage it. In recent years there has been a tendency for trainee archaeologists to take up specialisms, particularly in the environmental field, with the declared intention of increasing their prospects in the job market. Without wishing to claim professional exclusiveness, I see this trend at best as producing second-rate work and at worst as being downright misleading in a field like palynology.

IDENTIFICATION AND COUNTING

Again, identification is a job for the specialist. It is not difficult for the nonspecialist to recognize some of the more obvious types of pollen such as alder, heather or the Compositae, but he will fail to recognize more difficult taxa whose presence can be very significant even if they only occur in low numbers. Added to this are the problems of recognizing corroded, fragmented, damaged and contorted grains which are much more commonly met with in mineral sediments and deposits than they are in peat samples.

In the United States, Hall (1981) has investigated the distortion of pollen spectra where pollen is in poor condition. He found that taxa whose grains that are recognizable even when damaged or corroded are less of a problem than those which are not so readily recognized. In the latter category he includes oak (*Quercus*), elm (*Ulmus*), and grass pollen, all types of diagnostic importance. A point that I have not seen mentioned in the literature, but which can distract the novice, is that corrosion of smooth-surfaced grains can give them a superficial similarity to other grains, such as oak, which even in pristine condition have a rough surface.

Apart from mere identification, however, the inexperienced person can get into deep water over interpretation. Some archaeologists treat pollen grains themselves as though they were potsherds, useful for characterizing the cultural layers in which they are found. A pollen may be identified and given a name, but as already said, unless one is familiar with the plant itself and its ecology, adequate interpretation is impossible. The following true story illustrates my apprehension. One student, using the pollen analysis facilities in my laboratory, asked if he could come on a field excursion we were running. As we all disembarked from the Landrover I remarked that there was a good deal of *Carpinus* in the wood we had stopped at. In great excitement he jumped out, scrutinized the field layer (ground vegetation) and asked me to point out the *Carpinus* to him. *Carpinus* is, of course, a tree.

Identifying and then counting the grains is the most time-consuming part of pollen analysis and there seems to be no prospect of automating it. It is unlikely that the experienced eye of the specialist can ever be replaced by a machine; it is not just the size and shape of a grain which is diagnostic, but also the texture of the surface. As grains corrode and decay they can take on different textures which may still be characteristic. All such considerations have to be evaluated before an entry can be made on the counting sheet. In most cases, of course, there is no problem, but pursuing the occasional unknown or badly preserved grain can seriously slow the operation down.

The total number of grains which has to be counted from each sample to give a statistically valid representation of its pollen content varies according to the information required and to some extent according to the relative proportions of the different taxa. The subject is fully discussed by Moore and Webb (1978: Ch. 6). As a guide it may be taken that for a list of about 20 taxa, it is necessary to count 150–200 grains in order to get reliable percentages. This is adequate, for instance, if one is only concerned with the tree species, as was the case in the early days of pollen analysis. It is also adequate if one is only concerned with the main ecological dominants and the more important associated species in open environments, such as heathland, grassland, and farmland. But today pollen analysts sometimes identify a hundred or more taxa, in which case much higher counts, perhaps 1000 per sample, are needed to obtain adequate representation of the less abundant taxa in such a long list. Where one taxon is in great excess, as with, say *Calluna* (heather) in a heathland spectrum, or *Alnus* (alder) at a water-margin, its counts can be set aside, since its ecological significance is obvious, and statistical attention devoted to the less abundant taxa. Nevertheless, there are many occasions when the ecological significance even of a taxon whose pollen is present in excess is not obvious. In such a case it is wise to extend the counting until adequate representation of subordinate taxa is obtained.

There are a number of conventions in use for the construction of pollen diagrams. Moore and Webb (1978) give a useful account, though they omit the standardized symbols for the main taxa which are in general use in the Continental literature. However, a good account of this aspect can be found in Faegri and Iversen (1975). The preparation of diagrams will normally be the work of the pollen analyst himself, and in most cases the reading of the diagrams is straightforward, though the same is not necessarily true of their interpretation. Nevertheless, it is important in reading a diagram to establish what basis has been used for the calculation of the percentages. In the earlier literature this was usually the total of arboreal pollen (ΣAP). The reason was that the background forest was the standard against which variation of the non-tree pollen (NAP) could be seen. This works satisfactorily in circumstances when there was a well-wooded environment, though even here definitions vary as to what constitutes tree pollen. For instance in Britain hazel (*Corylus*) is conventionally regarded as a shrub and not a tree, even though it can reach a forest canopy. In Dutch diagrams birch (*Betula*) may be excluded and hazel included. There is also an increasing tendency to allow the interpretation to influence the basis of presentation. As mentioned above, alder (*Alnus*) pollen may occur in large quantities, particularly in aquatic deposits (Janssen 1959). As this doubtless represents the contribution of alder carr to the pollen rain, pollen analysts today frequently omit *Alnus* from the tree pollen total on the grounds that it is locally overrepresented. However, this begs the question of the status of *Alnus,* if any, in the composition of the original forest.

In a countryside which has been largely deforested it makes little sense to base the percentages on the tree pollen, for whatever tree pollen there is will be casual and may be the product of long-distance transport. In such cases it is usual to use the total pollen sum (ΣTP) as the basis of calculation. However, I know of one case, from the Highland Zone of Britain where the basis of calculation was the sum of Ericaceae, Gramineae and Cyperaceae, on the grounds that a moorland community was the general background ecological type. Unfortunately, the author neglected to state this on his diagrams. It is quite justifiable to use different bases of calculation to bring out particular points, but whatever basis is used should always be explicitly stated (e.g., Fig. 6.11 in Moore and Webb 1978).

The great majority of pollen diagrams published only show percentages (often called frequencies). Until relatively recently the actual amount of pollen in each sample (absolute pollen frequency, APF) was regarded as subject to too many variable influences, such as rate of decay, to be worth determining. In recent work on stratified deposits more attention is being paid to this, for various reasons that are not pertinent to this discussion. Much of the work on archaeological sites is an application of soil pollen analysis, and as we saw earlier the pollen frequencies can give an important clue to the long-term processes of pollen incorporation into the soil, and are therefore indispensable for full interpretation.

In the course of analysing pollen, other organic subfossils may be recognized and are sometimes so consistently present that they can be recorded alongside the pollen. An example of this which can be found in deposits and also in fresh water is the easily recognizable planktonic alga *Pediastrum*. Finely divided charcoal is commonly found in both stratified deposits (peats) and in soils; being elemental carbon it is indestructible by organic processes. But the most widespread inclusions of this sort are spores. These may come from fungi, mosses, liverworts and ferns. In many cases they are not identifiable, and even when they are, they may not be of great value for interpretation. For instance, in *Sphagnum* peats the abundance of *Sphagnum* spores may vary through a section, presumably depending on factors which promote spore production. It would not be valid to accept the spores themselves as an estimate of the relative abundance of the *Sphagnum* plants. For these and other reasons such spores, whilst often recorded, are not included in the pollen counts. With fern spores, however, there is a greater diversity of practice. For the most part ferns are plants that can form a characteristic component of the vegetation, comparable with, for example, the grasses or the Ericaceae, and their spore production is more nearly comparable with the pollen production of phylogenetically higher plants. However, the spores of ferns, some of which are more readily recognizable than those of lower groups, have one characteristic that affects their usefulness: they are very resistant to decay. In peat analysis they can be found at certain levels in very high percentages. These levels are generally thought to be the result of periods of greater microbiological activity, perhaps due to a drier phase which results in the overrepresentation of the resistant spores.

This factor applies equally, if not more so, to soils. As has been pointed out the lower samples of a soil profile may contain very high percentages of fern spores. In my work on soil pollen, which for reasons already explained largely refer to the acid soils, I decided to include fern spores in the sum from which percentages were calculated. My reason for this was that ferns such as bracken (*Pteridium*) could be a major ecological component of the vegetation and that high spore counts were not necessarily just the result of overrepresentation. Unlike peat analyses, where fern spores can be calculated as a percentage of a much greater sum (e.g., ΣAP), soil analyses are usually calculated on total pollen (ΣTP), so that by excluding what may be the contribution of a dominant taxon, the pollen spectrum would become meaningless. There is no single solution to this problem; one has to be aware of the different possible interpretations whichever method of recording is adopted.

2
WATERLOGGED SITES

INTRODUCTION

The majority of archaeological sites occur on freely draining terrain, but through the ages some habitats adjacent to bodies of water have been favoured, perhaps because they allow the exploitation of more than one kind of environment. They are also among the most productive habitats in terms of total organic matter. In mountainous regions, or in a landscape that is densely forested, the margins of bodies of water may offer the only habitat available for settlement. Whatever the reason, such situations are likely to be wet, and frequently they become waterlogged, either through climatic change or not uncommonly through changes in the hydrologic cycle brought about by man himself.

Whilst bodies of open water may be found in most regions with a climate suitable for human habitation, areas of high rainfall, particularly where impeded drainage may occur, can lead to boggy conditions with resultant waterlogging, and whilst archaeological sites are not common in peats (as opposed to *under* them, as will be discussed in Ch. 3), they do occur. Permanently high water tables may also produce waterlogging, as along some seaboards and watercourses. Occasionally there is a perched water table on an archaeological site, again producing anaerobic conditions. Artificial features such as wells, ditches or pits create localized points of waterlogging even on sites that have none of the hydrological features mentioned above, and are frequently the only places on such sites that permit the preservation of organic matter, including pollen.

Pollen is likely to be preserved wherever there is waterlogging all the year round, although one comes across deposits that quite inexplicably contain no pollen, even though the physical conditions seem suitable. Whilst it is primarily the anaerobic condition that reduces microbiological action, other factors may contribute. Sometimes pollen is less well preserved in a milieu of high pH than in an acid one, even though both are completely waterlogged. The presence of polyphenols in bog peats, or of tannins in some archaeological deposits (Biek 1963) on occasion also seem to exert a bacteriostatic influence.

It is important here to distinguish between permanent and periodic waterlogging. If a deposit is seasonally aerated, particularly if aeration coincides with the warmer summer months, microbiological activity can occur during that season. As a result any pollen originally present may be destroyed over a period of time. Even if pollen is preserved in a fully waterlogged deposit, it may not be possible to interpret it usefully. There is no means of telling the age or origin of pollen grains, so where pollen has been able to enter a deposit from various sources or at various times, it may not be possible to make any interpretation that is archaeologically or environmentally dependable. A number of these problems will addressed below in discussing features such as ditches as a source of pollen data (Ch. 4). However, partial interpretations can sometimes be usefully made even on inadequate data; for instance, if there is consistent occurrence of pollen of a taxon that is a strong ecological or chronological indicator this fact alone may be significant, whatever *mélange* of other pollen is present.

A great variety of archaeological sites, covering nearly all periods, occur in what today are waterlogged situations. This variety makes systematic presentation difficult; consequently somewhat arbitrary categories will be used, based on the conditions which produced the waterlogging, as described above. Our concern here is not so much to classify and cover the wide range of site types but to look at the application of pollen analysis which has been made to such sites.

SITES AT THE WATER'S EDGE

In this context we are immediately faced with the question of what constitutes a site. Bodies of water are habitual collectors of artifacts. If someone heaves an old bicycle over a bridge into a river, does this then create an archaeological site of the future? For present purposes I would exclude this. But if people habitually meet at the water's edge and as a result artifacts find their way into the water-laid deposits, there are grounds for regarding this as a site. A case in point is the occurrence of microliths in a mud deposit at Stump Cross on the Pennine Hills in Yorkshire (Walker 1957). The open water was apparently little more than a pond, but it laid down mud in which a number of microliths were found. Through pollen analysis of a section of the deposit, in which microliths were found *in situ,* Walker was able to establish that the occupation dated to the first half of the Atlantic period. If there were an occupation site here it would presumably be at the water's edge, but no evidence has yet been found of prolonged habitation in the vicinity. It was suggested that the site was merely a convenient resting place or a landmark visited from time to time. However, it did provide direct correlation between the cultural evidence and the environmental evidence as revealed by pollen analysis, giving a clear-cut date for the industry.

Another case of the dating of an industry, this time in the contrasting context of a mud deposit on the French Atlantic coast near Biarritz, was reported by Oldfield (1960). A severely eroded mud bed on the shore yielded artifacts attributable to the mesolithic Asturian culture. The pollen showed a clearance phase, possibly of anthropogenic origin, and this was dated by radiocarbon assay as 5100 ± 130 b.p.

No doubt in earlier times, as now, there were people who dwelt on or over the water rather than beside it, but I am unaware of any sites of this nature that have been investigated palynologically. At one time it was thought that the Swiss lake dwellings had been carried over the water on piles but there are alternative explanations of the archaeological evidence (see below). Crannogs might also be regarded as aquatic sites, though their existence was linked to the water margin. However, though palaeobotanical work has been done on seeds and other plant remains from crannogs, little has yet been published on their palynology.

For our present purpose it is not necessary to define the nature of a waterside site too closely. Fundamentally the mesolithic site of Star Carr is of the same nature as, say, a Swiss lake dwelling. Both are sites where man has created conditions at the waterside which enable him to live there either seasonally or permanently. They are both well described by the term used by Welten (1967)—Feuchtbodenwohnplätzen (i.e., dwelling places on damp soils).

At Star Carr (Clark 1954) a platform of timbers, brushwood, clay and stones was apparently constructed right at the edge of the lake, in a reedswamp community, the 'soil' of which was a detrital mud. The site was subsequently covered over by peat derived from the reed *Phragmites* which continued to grow through and over the occupation platform. Here the pollen analysis was carried out on a profile extending from the underlying mud, through the occupation level, and up into the overlying peat. It was a relatively uncomplicated stratigraphy, which made it possible to allocate this Maglemosian occupation to the Boreal period and so effectively to date it. This was, of course, before the days of routine radiocarbon dating. The local vegetation itself was directly reflected by the seeds and other macro remains present in the occupation level and the adjacent deposits, rather than by the pollen analyses.

Water's-edge sites, however, are not always as simple as Star Carr seems to be. Troels-Smith's interpretation of the early neolithic site at Muldbjerg is a case in point (Troels-Smith 1960). He established that the site was on a floating island; such islands frequently occur along the edges of reedswamp, where the plants form a floating mat of rhizomes which can be torn away in a storm and become free floating. Such floating mats are substantial enough to walk on and could serve as lakeside occupation sites. In this case drift mud collected beneath the island but above the basal calcareous mud of the lake, and clearly this drift mud must have been deposited subsequently to the island's breaking off. Troels-Smith used radiocarbon dating to elucidate this complexity, showing that the drift mud was of a comparable age to, or even older than, the neolithic occupation, and was markedly more recent than the base of the floating island which overlay it. This was proved by the discovery of matching sherds from the occupation layer and from the driftmud. Such a stratigraphy could hardly have been established from the pollen profile alone; it needed the additional evidence provided by the radiocarbon dates, and to some extent by the seed and other macroscopic remains present in the different layers. However, although the pollen alone did not permit a full elucidation of the stratigraphy, it did produce dramatic ecological evidence of early anthropogenic impact on the general forest cover. The impact was not only reflected in a reduction of the tree cover and the spread of grasses, but the weeds associated with early agricultural land use, particularly the ribwort plantain (*Plantago lanceolata*), occurred both in the occupation layer and in the underlying drift mud. Eventually the whole complex became embedded in the great Aamosen bog. The complete history had to be extracted from a series of peats and muds.

It becomes even more difficult to unravel the sequence of events when the occupation layer lies on and is covered by the same type of deposit. This is the case in a number of the Swiss lake dwelling sites giving rise to the so-called Pfahlbauproblem. Pollen analysis was applied to this type of site by Lüdi, Welten, and by Troels-Smith (who investigated Muldbjerg, described above). Typical of the sites investigated was an occupation layer embedded in calcareous clay, in a lakewater deposit, or in a mud, associated with the preserved remains of piles which had been driven into the clay (Troels-Smith 1954). From the stratigraphic evidence it is not possible to establish with certainty what the original situation was. Buildings of occupation had clearly been built on piles, but it cannot be conclusively established whether those buildings were built out over the water, or whether they were originally on a moist ground surface adjacent to the water's edge. Certainly the occupation layers subsequently became covered with a clay or mud similar to that on which they were deposited, but rises in the lake level may have occurred which could have produced this effect in either case. As at Muldbjerg the pollen analysis has had its greatest value in attributing the occupation layers to a particular stage of the Postglacial. In demonstrating the effect of early neolithic occupation of the surrounding landscape. Troels-Smith was able to show an impact remarkably similar to Muldbjerg, but involving different tree species, in Denmark and Switzerland at about the same time. The pollen record, however, was able to contribute little to the elucidation of the original state of the site and the changes which subsequently took place in it. Again macro remains gave important evidence of specific culture plants, so establishing more certainly than the pollen could the local use made of such species in the contemporary economy. Another form of construction in Swiss lakeside sites in the Early Neolithic was the so-called Packwerkbau. Here the foundations were not piles but layers of crossed timbers. There is an argument in this case too as to whether the timbers rested on the lake bottom or were set on the damp soil marginal to the lake. At the site of Niederwil, studied by Waterbolk and van Zeist (1967) and later by van Zeist and Casparie (1974), there was some indication that the occupation horizon lay at the interface between gyttja and peat. The gyttja would have been laid down underwater and van Zeist and Casparie believe that it could have been exposed later by a drop in the lake level perhaps brought about by the neolithic people themselves; in other words that the dwelling site was terrestrial and not aquatic. Subsequently the site was buried beneath sedge peat as the whole lake basin piled up with peat. Botanical investigation revealed the presence both of seeds and pollen of crop plants and of weeds at the occupation level of the site itself; the overlying peat contained no indication of farming, suggesting that peat formation did not take place until the settlement had been abandoned. As with the other early neolithic sites already referred to above, there was clear evidence of forest clearance and the spread of shrub vegetation at the forest edge. For reference purposes a long pollen profile was obtained from the deepest part of the lake basin.

Whilst it is not suggested that prehistoric occupation of such habitats always occurred on the lake shore rather than over the open water, these damp soils at the water's edge were clearly a commonly used human habitat. One example from a context totally different from those described above is a wooden structure, again of neolithic date, embedded in fen peat beneath the later peat of a raised bog in the English Lake District (Oldfield 1967). The structure consisted of broad wooden planks overlying a foundation of twigs and branches and it was demonstrated that the structure had been set on the junction between mineral soil (a thin smear of soliflucted drift overlying carboniferous limestone) and fen peat indicative of open reedswamp conditions.

An interesting association of archaeological remains and peat was explored by Tomalin and Scaife (1980). The peat deposit gave a very complete pollen sequence in which evidence of human activity could be detected. In view of this a programme of field walking was carried out up to 100 m from the mire; it yielded nearly 300 struck flints dating from the Mesolithic to the Late

Neolithic. Suggested correlations were made with the pollen evidence. Though the archaeological material was abundant and in places concentrated, no circumscribed site in the usual sense of the word was identified and investigated, and certainly there was no site in the waterlogged deposits themselves.

In addition to ponds and lakes, permanent watercourses can provide sites for waterside settlement. At a river or stream level the water table comes to the surface, so one may expect that conditions for the preservation of organic matter will be favourable. Nevertheless, a river's-edge site may not be suitable for settlement or occupation as river levels vary seasonally. Settlements in earlier times avoided flood plains, a lesson we could usefully relearn in some of our modern urban developments. Sites associated with a permanently as opposed to a seasonally high water table will be dealt with below under another heading, but here it may be observed that river courses can yield archaeological/environmental sites in areas which are otherwise virtually sterile. A case in point is the Acheulian site at Kalambo Falls on the boundary between Zambia and Tanzania. In a region of long dry seasons and high temperatures ancient organic remains are very scarce, but in the Kalambo River gorge the permanent wetness had allowed the preservation of plant material, including pollen, in association with an Acheulian floor. Not only does such evidence add an important dimension to the archaeological findings, but it has its own intrinsic scientific value in relation to past ecological and climatic history. Van Zinderen Bakker (1969) points out that "nearly all the trees of the forests of Africa are insect pollinated . . . consequently . . . arboreal pollen is greatly under-represented in Africa as compared with non-arboreal pollen." Nevertheless, even on the limited material available van Zinderen Bakker was able to obtain meaningful pollen analyses for a series of discrete samples from bands of organic matter preserved below the water table. These spanned a period of some 50,000 years and gave evidence of climatic conditions, particularly temperature, at various points over this time scale.

It was, incidentally, an African swamp deposit that gave the highest absolute pollen frequency that I have come across; a sample from the Lake Bangweulu swamp gave a total frequency of over 15 million grains per gram of deposit. Of this quantity 80% was of one taxon, *Syzygium*, a wind-pollinated swamp forest tree (Lawton 1959).

Even a river alluvium can sometimes yield pollen, though it is not easy to be sure where the pollen originated. In a wadi in Jordan, a sample from the Younger fill, which contained Roman and medieval artifacts, yielded a pollen spectrum dominated by Chenopodiaceae and containing few tree taxa (Vita-Finzi and Dimbleby 1971), but the spectrum could not be correlated reliably with the conditions controlling wadi-flow.

Recent work in Britain is suggesting that an important factor in the build-up of valley alluvium may be erosion from the slopes of the catchment. Shotton (1978), using sediment analysis, has concluded that in the Severn Valley there was an influx associated with Late Bronze Age farming, and Scaife and Burrin (1983), using both sediment analysis and pollen analysis, conclude that in a Sussex catchment sediment built up at times of forest clearance.

A similar uncertainty about the source of pollen limited the interpretation of a pollen profile taken from river sediments outside the North Gate of Roman Verulamium at St. Albans (Dimbleby 1978b). Here there was no difficulty in obtaining pollen counts, but it was not possible to say with certainty whether the pollen was local to the sampling point or whether it had come from a wider area. Recent work in the palynological field is indicating that a great deal of pollen in lacustrine or fluviatile deposits may be brought down with the river sediments. If this applied to the Verulamium analysis, as seems probable, the results may say more about land use upstream than about what was going up near the sampling point. In this particular case the pollen assemblage was strongly agricultural with only a minimal representation of trees and shrubs. The landscape was apparently open then as it is now.

SITES IN PEAT

It is virtually impossible to classify wetland sites in such a way that the categories are completely distinct. Certainly, this applies to the distinction between the last category and this one: where there are ecological transitions there are bound to be sites that straddle the dividing line.

An example of such a site is the Magdalenian springhead site at Schussenquelle, on the watershed north of Lake Constance in southern West Germany (Lang 1962). Though the basal parent material of the site was clay overlying moraine, the occupation level itself was in a 75 cm layer of peat, produced by the moss *Drepanocladus aduncus* and radiocarbon dated at 13,350 b.c. This species of moss is characteristic of semiaquatic habitats, particularly where conditions are neutral or slightly basic. The pollen diagram showed that this occupation occurred in the earliest phase of the Older Dryas period (Zone Ia), before the postglacial spread of forest trees had begun. As is usual in peat deposits, macroscopic remains of plants were also present and confirmed the pollen analyses, giving additional evidence about species present locally.

Archaeological sites centred on springs can be a valuable source of environmental evidence. Simmons (1969), for instance, was able to establish the environment of mesolithic man on the North York Moors and to demonstrate that even at this stage of cultural development the impact of the occupation on the local ecology was detectable.

As fen peat normally forms under a water surface or in reedswamp, sites associated with it will as a rule be most appropriately placed in the previous category: sites of the water's edge. On the other hand, sites which are embedded in peat, both above and below, are more likely to be in a raised bog or in blanket peat. In these cases the peat is saturated, not by base-rich groundwater, but by rainwater (ombrogenous water), and consequently is acid. Obviously an acid bog is not conducive to settlement—it holds few of the necessities of life, but an abundance of disadvantages. Archaeological sites may be found, but they are rarely settlement sites.

One such case was a remarkable timber structure, surrounded by a ring of stones, set in the peat of the bog at Bargeroosterveld in Drenthe province, Holland (Waterbolk and van Zeist 1961). The timber work was well enough preserved to permit a reconstruction which took the form of a rectangular open-sided, ornamented, but unroofed structure, with no known parallels in the archaeological records. No pottery and no artifacts of metal or stone were found associated with it. A pollen sequence was established through the bog and on this basis the age of the structure was estimated to be about 1000 b.c. Subsequent radiocarbon dating confirmed that it belonged to the Middle Bronze age. The structure had no apparent functional purpose and it was concluded that it was a sanctuary. Ritual burials of bronze hoards, corpses and other objects in bogs were well attested, showing that the bog environment held some interest beyond that of settlement.

A great deal of wetland research has been carried out using pollen analysis as a main tool. In Britain the Fen country of East Anglia has been extensively studied for half a century, and it has provided a firm environmental background for archaeology, with applications far beyond the local area. In particular the pattern of forest change, and introduction and development of agriculture, have been clearly shown. An excellent general account of this work is given by Sir Harry Godwin (1978) in his book *Fenland: its ancient past and uncertain future*.

Another extensive area of wetlands, the Somerset Levels, has also been intensively studied, but in this case archaeology has been the focus of study. Commercial peat digging has been in progress for many years and as a result of this and of field reconnaissance, it has been established that this large level area of boggy ground is traversed by a series of prehistoric tracks, now known to span the fourth to the first millennia B.C. in age (Orme 1982: Fig. 11). At intervals the monotony of the levels is broken by low hills and it is clear that some of the tracks connect the hills, which are in effect islands protruding through the flat bog. Others follow less obvious courses, perhaps leading to open water or other situations of interest to prehistoric man. Though they vary in detail of construction, these tracks consist of timber on poles laid over the peat, sometimes on a brushwood base, and pegged into the underlying peat. Also scattered through the levels numerous small wood or brushwood platforms have been found, not apparently connected with the trackways.

The picture that emerges is that such an extensive area of bog was not shunned by man but was regularly visited and crossed. Over a long period of time many archaeological finds have been made, not indicating an occupation, but clear evidence of the passage of man. However, at the end of the last century two settlements were located out on the bog, one near Glastonbury and the other near Meare. Parallels with the Swiss lake dwellings were quickly realized, and the Glastonbury site was extensively excavated. Some early excavation also took place at Meare, but enough of the site was left for new excavation in 1978 and 1979, in which not only modern archaeological techniques, but also methods of environmental research not available at the time of the early excavations, were employed (Caseldine 1983b). Peat underlies the site, but there are abundant artifacts, charcoal, and macro remains of domestic plants and animals. The site has been dated to the Iron Age and there is evidence that it was subject to flooding both during its active life and after abandonment. Though puzzling questions obviously arise from this unusual settlement habitat, there are indications that it was occupied intermittently, perhaps in years or in seasons when the water table was lower than usual.

I shall not pursue the account of the Somerset Levels project here, for our concern must be with pollen analysis, and the research being carried out there now goes far beyond this one technique. In the early days pollen analysis was used to establish dating, but this has now been superseded by radiocarbon. On the other hand the pollen data have given invaluable evidence about the sequences of vegetation change, some of them clearly the result of human activity in prehistoric times. One of the advantages of a peat medium is that it preserves organic matter generally, and full advantage of this has been taken by Professor John Coles and his team. Botanical material other than pollen has provided a great deal of ecological detail, and this has been augmented by studies on the faunal remains of insects and mites. A detailed species analysis has been made on the timber and wood of the trackways, together with tree-ring studies on as many species as possible. Microbiological work on the wood has added to the picture, and it has even been possible to show that the timber for the trackways was produced by deliberate coppicing. The *Somerset Levels Papers*, a periodical that has appeared annually since 1975, presents the archaeological and scientific evidence in detail, with an abundance of illustrations of this remarkable work.

HIGH WATER TABLES

Many of the sites already referred to are, of course, associated with high water tables but in these cases the waterlogging has resulted in deposits such as peat or gyttja which have brought about preservation and made it possible to apply pollen analysis to the stratified deposits. There are circumstances, however, where settlement has taken place on land which has a high water table but which has not developed such deposits. Good examples of such sites are the *wierden* or *wurten*, mounds built up to allow settlement on wet land subject to flood. These occur in large numbers along the low-lying German North Sea coast and the estuaries leading into it, and their dates encompass the first fifteen centuries of our era. The mounds themselves, which have often been enlarged and built up many times, are constructed of layers of stable manure, clay and turf and provide perfect conditions for the preservation of organic matter. A most detailed botanical investigation was made by Körber-Grohne (1967, 1981) on one of these mounds, the Feddersen Wierde, in which she concentrated largely on the macroscopic remains of the plants, relating them to the agricultural practices as well as to the local ecological conditions. Pollen analysis was of limited use but was used to amplify the other botanical evidence. A similar investigation, also concentrating on macro remains, but making use of pollen analysis as well, was made by Behre (1976) on a wierde called Elisenhof.

Behre also carried out botanical research into the Viking settlement of Haithabu, situated near the modern Schleswig, northwest of Kiel on the German Baltic coast (Behre 1969). In order to put the seed and wood analyses into a wider ecological setting Behre carried out pollen analysis on a number of moss polsters found in the excavation. These revealed a wide range of pollen spectra, some clearly coming from woodland environments. It was concluded that some at least of the moss polsters had been deliberately brought into the town perhaps for packing or caulking, and that they therefore gave a glimpse of habitats outside the settlement.

Viking settlements are commonly situated at the head of navigable estuaries and consequently the old levels are often waterlogged. A good example, which has been the subject of detailed multidisciplinary investigation for some years, is in the city of York. Here there has been magnificent preservation of organic remains; however, as with the other examples given under this heading, it is the analysis of the macro remains rather than pollen which most usefully gives the environmental dimension to the archaeological discoveries.

With the increasing archaeological activity in many of our ancient towns and cities it is frequently found that there was a close association between urban developments and waterfronts; the watercourses themselves have often long been obliterated. As water tables tend to be high in such situations, the preservation of organic matter may be very good. In a few cases pollen analysis has been used, though the results may prove to be of more cultural than environmental value (e.g., Ayers and Murphy 1983; Hall *et al.* 1983).

The Roman frontier-post of Vindolanda, about midway along Hadrian's

Wall, provides a quite different example of the influence of a water table. Here pre-Hadrianic levels were discovered by accident in the course of digging a drain through what was thought to be basal glacial clay (Birley 1977:104). Wonderfully preserved organic remains, including many artifacts, lay beneath this clay layer. There had clearly been occupation, including a leather-making workshop. When a trench was opened, water poured in, indicating that the apparent water table was in fact a perched one, created by the layer of clay and other impervious deposits overlying the pre-Hadrianic levels. This effectively kept these levels anaerobic, and the presence of a great deal of bracken, dung, and urine, apparently associated with the leather industry, had no doubt contributed to the virtual absence of microbiological activity. It has been shown that pollen is abundant and in good condition in these deposits, but detailed pollen analysis has not yet been carried out.

WELLS, PITS, AND SUMPS

Even on freely-draining sites waterlogged conditions may occur in features which have been cut into the ground. Wells are an obvious example, being deliberately sunk until groundwater is reached. Some pits, which were clearly not designed to be waterlogged, may have become so inadvertently, or in some cases because there has been a subsequent rise in the water table. Other features may have been deliberately cut in order to collect excess water from the site; that is, they are sumps. All these may give excellent preservation of organic matter. Noel Hume (1976) describes the contents of a well at eighteenth century Williamsburg, Virginia, and illustrates perfectly preserved leaves of holly and box that were still green when excavated. This greenness is a not uncommon feature of anaerobic deposits. I have recorded green moss from the Bronze Age shaft at Wilsford, near Stonehenge, but the colour rapidly faded on exposure. A number of botanical investigations of plant remains from wells have been undertaken, but they have not often included pollen analyses. The Wilsford Shaft, however, was a case where palynology was carried out on two samples (a third sample was almost devoid of pollen; Table 1) but the results were consistent with each other and reflected an arable and pastoral agriculture in the chalk landscape. This conclusion was confirmed by the mosses and other plant remains, as well as the abundant insect remains (Osborne 1969) contained in the deposit. Though archaeologists disagree about the purpose of this shaft (from which the remains of a stave bucket and bast rope were recovered) it does represent a unique micro-environment in the chalk landscape of Salisbury Plain, giving unparalleled evidence about the environment in the Bronze Age. Unfortunately the definitive report on this site has not yet been published.

Table 1

POLLEN ANALYSIS OF TWO SAMPLES FROM WILSFORD SHAFT (PERCENTAGES OF TOTAL POLLEN PLUS FERN SPORES)

	Sample no.			Sample no.	
	9	230		9	230
Alnus	0.4	0.5	Cruciferae	0.4	—
Betula	0.2	—	*Cuscuta*	0.1	0.5
Fagus	0.1	0.5	Cyperaceae	0.7	—
Fraxinus	0.3	—	*Helianthemum*	0.5	0.5
Pinus	0.1	—	Papilionaceae	0.2	—
Quercus	2.0	2.3	*Plantago lanceolata*	11.2	6.8
Ulmus	0.1	0.5	*P. major/media*	1.2	1.4
Corylus	1.5	—	Ranunculaceae	2.5	—
Hedera	0.3	0.5	Rosaceae	0.5	—
Prunus	0.1	—	*Potentilla*	—	0.5
Salix	0.1	0.5	*Poterium*	2.2	1.4
			Rubiaceae	1.3	—
Calluna	0.1	—	*Rumex*	0.3	0.5
Gramineae	31.6	48.2	*Succisa*	0.1	—
Cerealia	3.4	4.1	Umbelliferae	1.6	1.4
Boraginaceae	0.2	—	*Urtica*	0.2	—
Caryophyllaceae	0.6	0.5			
Chenopodiaceae	2.3	2.7	Varia	3.6	1.8
Circaea	0.1	—			
Compositae			*Dryopteris* type	11.2	5.0
Artemisia	0.7	3.6	*Polypodium*	—	+
Centaurea nigra	0.3	0.9	*Pteridium*	9.9	9.0
Liguliflorae	5.6	4.5			
Tubuliflorae	2.0	2.7			

Sample No.	9	230
Total pollen plus fern spores	992	222
NAP/AP%	2300	2288

Deep-dug features in river valleys are also susceptible to waterlogging for obvious reasons, and on occasion this has been turned to good use. In his investigation of Iron Age and Roman settlements at Farmoor on the terraces and floodplains of the Upper Thames, Robinson (1978; Lambrick and Robinson 1979) took bulk samples of unstratified waterlogged deposits, which were then subsampled for pollen analysis. The analyses were consistent in showing only traces of tree pollen but abundant evidence of grassland, probably hay meadow, and both pastoral and arable agriculture. Two of the samples came from Iron Age sumps or gullies, two from Roman wells, and one from a peat layer of Roman age.

The pollen spectra from the Roman wells show a remarkable similarity to that obtained from a Roman well at Pforzheim (Dauber *et al.* 1955). This well, one of five in the area examined archaeologically, showed only 5.2% of the total pollen to be tree pollen, despite the fact that the wells contained a great mass of wood, worked and unworked, from over a dozen species.

A somewhat different type of feature, namely stone-built sewer system at York, also of Roman date, produced a deposit that yielded a pollen spectrum (Buckland 1976). The spectrum suggested a rural landscape, with woodland and heath elements, quite at variance with what would be expected in the centre of a Roman town. One can speculate about how the pollen got into the sewer; it may have come with the water supply, or it may have found its way from a pre-Roman land surface through which the sewer had been laid. In the absence of any positive indication of the source of the pollen this must be put down as one of those cases where pollen is present but its analysis cannot be usefully interpreted in terms of environmental archaeology.

3
OLD LAND SURFACES

INTRODUCTION

At one time archaeologists often used the term 'the natural' to describe the unmodified 'soil' beneath an earthwork, and one of the means of recognition of 'the natural' was that it was usually yellow or orange in colour. Fortunately, the increased awareness of environmental factors has now largely eliminated this terminology, and it is recognized that the so-called 'natural' was generally the C horizon of a buried soil whose true surface, on which man was operating, was at a higher level and had in the past been trowelled away. It has become the practice to refer to the 'old land surface' (if it can be recognized), though there is still some confusion about what constitutes a surface. Not infrequently it is described as a humus-rich layer perhaps 10 or 20 cm thick—hardly a true surface. As was shown in Chapter 1, pollen analysis can often locate a buried surface with precision (as exact as the sampling interval allows), even in sections in which a soil surface cannot be recognized visually.

At the meeting of the British Association for the Advancement of Science in 1968 a discussion group gave exhaustive consideration to buried soils and their implications. The group was made up of scientists from a number of disciplines and a synopsis of its findings was published the following year (Dimbleby and Speight 1969), which will provide a useful background to this chapter.

There is a wide variety of ways in which an old land surface can become buried and in what follows the most important of these will be discussed, with some examples (often those I have worked on myself) of the application of pollen anaysis to sites so buried.

BURIAL BY VOLCANIC ASH

It is a frequent problem in studies of buried soils that there is pollen in both the soil itself and the covering material. As we have seen the actual *quantity* of pollen rather than the relative frequencies of the different taxa may be the indication of the level of the old surface. We will start with a simple situation in which this question does not arise—burial by volcanic deposit. There are many instances in different parts of the world of archaeological sites being buried by volcanic outfall. One well-known example is the burial of the Minoan sites of Crete, such as Knossos, by ash from the volcanic eruption on Santorini. In such circumstances one would not normally think of applying pollen analysis to the buried soil, and in some cases, as where this soil is reddened by heat, it would obviously be pointless. But at Pompeii Professor Wilhelmina Jashemski, who had been excavating the gardens of the destroyed town, had the idea of looking for pollen in the garden soil. It turned out that there was pollen present in small amounts, which cast a little light upon the contemporary garden plants (Dimbleby 1976) but also upon the nature of the covering deposit. This consisted of lapilli, which would be pollen-free, and which strongly contrasted in colour with the garden soil beneath. Moreover, as the depth of the lapilli was at least 6m, there is clearly no question of pollen having found its way into the buried soil from the modern surface. The fact that polien was present must indicate that the volcanic outfall was not very hot at this point (though there may have been toxic gases), and this is corroborated by the fact that tree trunks and roots were not charred but had decayed *in situ,* leaving large cavities in the lapillar cover and in the buried soil. The archaeological significance of this pollen work is not as great as might have been hoped, but garden soils are not a propitious medium for pollen analysis under the best of conditions. However, it was established that olives were being grown, a fact not previously known, and one that could be correlated with other archaeo-horticultural evidence found in the buried soil. The buried soils of these gardens have been made to yield a remarkable amount of information about the gardens, showing parallels with present-day techniques. The work has been beautifully presented and illustrated by Professor Jashemski in her important book *The gardens of Pompeii* (1979).

BURIAL BY AEOLIAN SEDIMENTS

Blowing sand is an active form of mass movement of soil which has frequently covered archaeological sites. This is particularly true in coastal sand dune systems; some well-known examples are Skara Brae in the Orkneys (Neolithic), Hengistbury Head in Hampshire (Mesolithic), and Gwithian, Cornwall (Bronze Age). Sand blow is not, however, confined to coastal areas; West Stow in Norfolk is a good example of a multi-age site covered by sand rendered unstable by deforestation and overintensive agricultural use.

Whilst such situations are of particular value archaeologically, for the sand can offer some protection against disturbance to the structures and artifacts, only rarely has the old land surface beneath blown sand been examined palynologically. There are two records of such investigations in the Isles of Scilly (Dimbleby 1976–1977; Dimbleby *et al.* 1981). Neither, however, was associated with archaeological material, though both showed evidence of prehistoric land use and one established beyond question that those islands had once been wooded.

The fortuitous finding of a hoard of fifteenth century coins in the sand dunes at Glenluce, Wigtownshire, led to an excavation which revealed an old land surface beneath the dunes (Jope and Jope 1959). In fact there were two superimposed surfaces, both horizontal, showing that although there was apparently sand movement at the end of the fifteenth century, the present unstable dune system had only developed subsequently. The two buried surfaces were examined palynologically and both revealed a grass-heath type of vegetation with few trees or shrubs, but whereas in the older (deeper) surface the grass pollen exceeded in quantity that of heather, in the younger surface the roles were reversed. The reversal is a possible indication of a pattern of land use, particularly the use of fire, which was progressively acidifying the soil. It was not possible to say with certainty what the relative ages of the two surfaces were, but a difference of 50–100 years would be consonant with the archaeological and numismatic indications.

In the Netherlands and in Belgium one or two brief investigations have been made of archaeological sites beneath dune systems. Vermeersch *et al.* (1973, 1974) have studied mesolithic sites within or beneath sand dunes, but it proved difficult to tie the pollen data in with the archaeological material. Havinga (1964) studied the pollen in a sand-covered Iron Age site in Holland. Open sites, in areas of sandy soils, may show evidence of sand blow, but these are not buried beneath a massive overburden of dune sand and have to be interpreted differently. They will be dealt with in detail in Chapter 5.

Aeolian dunes may, of course, form in material other than sand. Loess, being itself an aeolian deposit, is liable to be reworked by the wind when exposed under dry conditions. In general loess, having a relatively high pH, does not favour the preservation of pollen, but on occasion it can give positive results. One such case was the site of Vadastra in the basin of the Lower Danube. Here Aurignacian artifacts occurred on a loess dune above the river terrace, which was then progressively covered by loess deposition. The palaeolithic occupation ceased, but near the present surface neolithic industries appeared. Leroi-Gourhan *et al.* (1967) carried out pollen analyses on samples from each recognizable stratum; these showed climatic differences in the various stages. No attempt was made to carry out serial sampling and no surfaces were investigated; had this been done the apparent discrepancies between the pollen evidence and that from molluscs in the same layers might have been more easily explicable.

BURIAL BY MARINE AND FRESHWATER TRANSGRESSIONS

Another coastal process which can bury old land surfaces is a transgression of the sea, brought about by a rise in sea level or a drop in the land base—in areas where isostasy and eustasy operate, commonly a combination of the two. Such a transgression may lay down marine clays over extensive areas, as in the Fenland of East Anglia (Godwin 1978), the Somerset Levels, or along the Lancashire coast. Many examples are known of artifacts preserved under such clay deposits, though there are few recorded instances of archaeological sites being covered in this way, and fewer still that have been investigated palynologically. Estuaries are the most likely situation for burial in this way, where a river channel may funnel water to exceptionally high levels. This was the case above the Welsh coast of the Severn estuary, when the Roman quay at Caerleon was buried by a deposit laid down by such a flood, probably in A.D. 1606 (Boon 1980). It was not appropriate to investigate this palynologically.

Flooding by freshwater may also occur along coasts as well as over flood plains inland. Louwe Kooijmans (1980) describes the intercalation of freshwater peats and clays, the latter carrying prehistoric occupation, at Hazendonk in the Rhine estuary, and he carried out pollen analyses of these cultural levels. Archaeological sites on flood plains have been excavated, but the application of pollen analysis has been confined to the fillings of ditches, wells, and so forth (e.g., C. Smith 1979; Lambrick and Robinson 1979). Old land surfaces may be difficult to recognize beneath alluvium (see below) and in any case conditions of pollen preservation are often not ideal. Fluctuating water tables and deposits of high pH give favourable conditions for microbiological activity, and such deposits are also liable to be worked by earthworms.

BURIAL BY BLANKET PEAT

Burial of an old land surface by blanket peat might be thought to be more appropriate to Chapter 2, for the formation of peat implies waterlogged conditions. However, I am deliberately including it here in order to make the point that the old land surface itself may not have been waterlogged before the peat was formed. In fact it is often found that the soil beneath the peat is not waterlogged and the soil profile is that of a normal aerobic soil. Indeed, in Ireland or in the Highland Zone of Great Britain there are many examples of early occupation, ranging from the Mesolithic onwards, now covered by several metres of peat.

The chronological and ecological context of some peat-covered sites have been investigated indirectly by pollen analysis of adjacent peat deposits, but in a few cases the actual sites themselves have been subjected to pollen analysis. Two mesolithic examples of this were at Cook's Study, near Holmfirth in the Pennines (Dimbleby 1969), and White Gill on the watershed of the Cleveland Hills in North Yorkshire (Dimbleby 1962a:111–113). In both cases microliths were found on the mineral surface of a soil buried beneath blanket peat, and in both cases the pollen in the mineral soil indicated a woodland environment, whilst the overlying organic layers showed relatively open conditions. There was probably a chronological hiatus between the pollen record in the mineral soils and that in the overlying organic matter, and there were indications that man himself was bringing about ecological change within the mesolithic period. There is also evidence that on the North York Moors, at least, forest regenerated itself after the mesolithic occupation and that the landscape was not finally opened up until the Bronze Age.

In this area there is little direct evidence of neolithic occupation, but in other moorland areas there was a definite neolithic presence. Such a case has been studied by Pilcher (1969). The complex neolithic site at Beaghmore, County Tyrone, Northern Ireland, provided several distinct contexts for pollen analysis, including a lake basin, a land surface covered by blanket peat, and the land surface covered by a mound. These sources provided clear evidence of change during and after the neolithic period. The mound was built after forest clearance and a period of extensive agriculture. Radiocarbon dates from the base of the peat showed a wide range; this may indicate that peat formation began at different times at different points. On the other hand, the fact that the pollen analyses suggested a break in the sequence between the mineral soil and the overlying blanket peat could be explained not only by peat starting to form at different times, but by the destruction of peat after it had formed. An obvious agent for this is fire, which could have occurred repeatedly, so increasing the chronological and palynological unconformity at this juncture.

It is not appropriate here to discuss the theory that the origin of blanket peat is linked with human activity; a useful summary of the evidence is given by A. G. Smith in Simmons and Tooley (1981:109). It is often assumed that blanket peat will only develop in periods of wetter climate, but this is not necessarily so. The development of surface waterlogging is not a simple reflection of climate but involves factors such as soil structure and the nature of surface humus. This is well illustrated by the sequence of events established at the neolithic site of Goodland, lying beneath blanket peat in County Antrim (Case *et al.* 1969). Here it was possible to suggest a sequence as follows:

1. Clearance of forest from a brown podzolic soil.
2. Cultivation of a soil with an active earthworm population.

3. Pasture, with progressive soil acidification leading to:
4. The disappearance of the earthworms and the development of a leached soil with a thin iron pan: such a soil would have a poorer structure in the A horizon than the previous earthworm-mixed soil.
5. This poorer structure leading to the development of surface wetness and the replacement of grass by rushes (*Juncus* spp.).
6. This vegetation producing a dispersed colloidal humus which impedes drainage, leading to the onset of blanket peat growth which radiocarbon dating shows to have started within the Neolithic.

It should be noted that the Subboreal period, in which the Neolithic falls, was long held to have had a drier climate than the periods before or after, though in the oceanic climate of Northern Ireland dryness is a relative term.

The sequence outlined above was established by a combination of research approaches: pollen analysis, seed analysis, pedological evaluation, radiocarbon dating, and, of course, archaeology itself. Pollen analysis will usually show whether there is a major hiatus between the buried soil and the overlying peat. If there is, the site can be studied on an *ad hoc* basis like any other buried site. Impedance to drainage means that water movement down into the buried soil is probably almost static, so that contamination with later pollen can be discounted, as the clear-cut boundaries prove. Where there is no such uniformity the site itself can still be investigated in the same way but it also offers the possibility of establishing ongoing landscape changes originating from that occupation.

BURIAL BY HILLWASH

The movement of soil downhill is of course a natural process. In co climates it may take the form of solifluction in which the whole soil mantle slips down over a base of frozen subsoil. In warmer climates, especially where there is either a marked wet season or a tendency to sudden storms even in the dry season, soil particles will be moved downhill under the influence of gravity. The rate at which this takes place depends on the steepness of the slope and on the incidence of rain, but it is largely controlled by the cover of vegetation. Forest cover reduces the effect to the minimum, but even a grass sward exerts effective control. If, however, the plant cover becomes disrupted, and particularly if the humic surface horizon of the soil is broken, the effect of rainfall on soil movement is much increased. Depending on circumstances gullying may occur, but more generally there develops a process of sheet erosion, that is, the mass movement of the upper layers of the soil downhill. Sometimes this is referred to as accelerated erosion, but the factors which cause erosion are present all the time; their impact is reduced by the plant cover, as described above. The erosion is only accelerated because the brakes are taken off.

Wherever man interferes with the plant cover of the earth, he allows the erosive power of the local environmental factors to increase their effect. This does not necessarily have to involve agriculture. Even in the Mesolithic in Britain it can be shown that mineral soil has been washed into basin sites at a time when the tree cover was being reduced, probably through the use of fire (Simmons *et al.* 1975). There is no example yet of a mesolithic soil being buried in this way, but when we turn to the Neolithic such cases are found. Evans and Valentine (1974) described a soil buried beneath hillwash on the scarp of the Chiltern Hills. A radiocarbon date of 1996 ± 220 b.c. puts the episode in the Neolithic period, and the analysis of molluscs from the soil clearly demonstrates an ecological sequence from woodland, through a clearance phase, to tillage. Nevertheless, soils buried by hillwash in the neolithic period are not common, at least in Britain; presumably the extent of arable land was limited and probably it occurred in discrete blocks. However, when we turn to the Late Bronze Age and the Iron Age, when agriculture was much more extensive and intensive, there is widespread evidence of massive soil movement. Work by Bell (1981) has shown deep deposits in dry valleys in the chalk of southern England, with earlier archaeological levels covered by later hillwash. It had been suggested (Godwin 1967) that chalk was too porous a bedrock to allow erosion, but this is clearly a false assumption. Though such hillwash is certainly widespread, it has only been investigated in detail in calcareous soils, which are not suitable for pollen analysis. Bell did find old land surfaces which had been decalcified to such an extent that they contained pollen, but in these soils palynology could not yield as much information as mollusc analyses.

Where hillwash has been active the old land surface may be difficult to detect visually, if it is there at all. In some places Bell found that the original soil seemed to be missing, having presumably been eroded long before the site became covered with material from upslope. Even where it is present, as on a site of low gradient, it may be virtually unrecognizable. If the old land surface has been cultivated and then becomes covered over by topsoil brought down from upslope, the colour and texture of the plough soil *in situ* and that which has been deposited over it may be identical. In other circumstances, such as where the old land surface was under pasture, an earthworm-sorted stone-free zone may have developed which gives a visible textural contrast. In either case, of course, any pollen in the soil will have become mixed and therefore difficult if not impossible to interpret. In calcareous soil where decomposition of pollen is rapid, it can be argued that any pollen present must be all of the same age (see Ch. 1) that is, contemporary within a year or two of the last exposure of the surface (Dimbleby and Evans 1974).

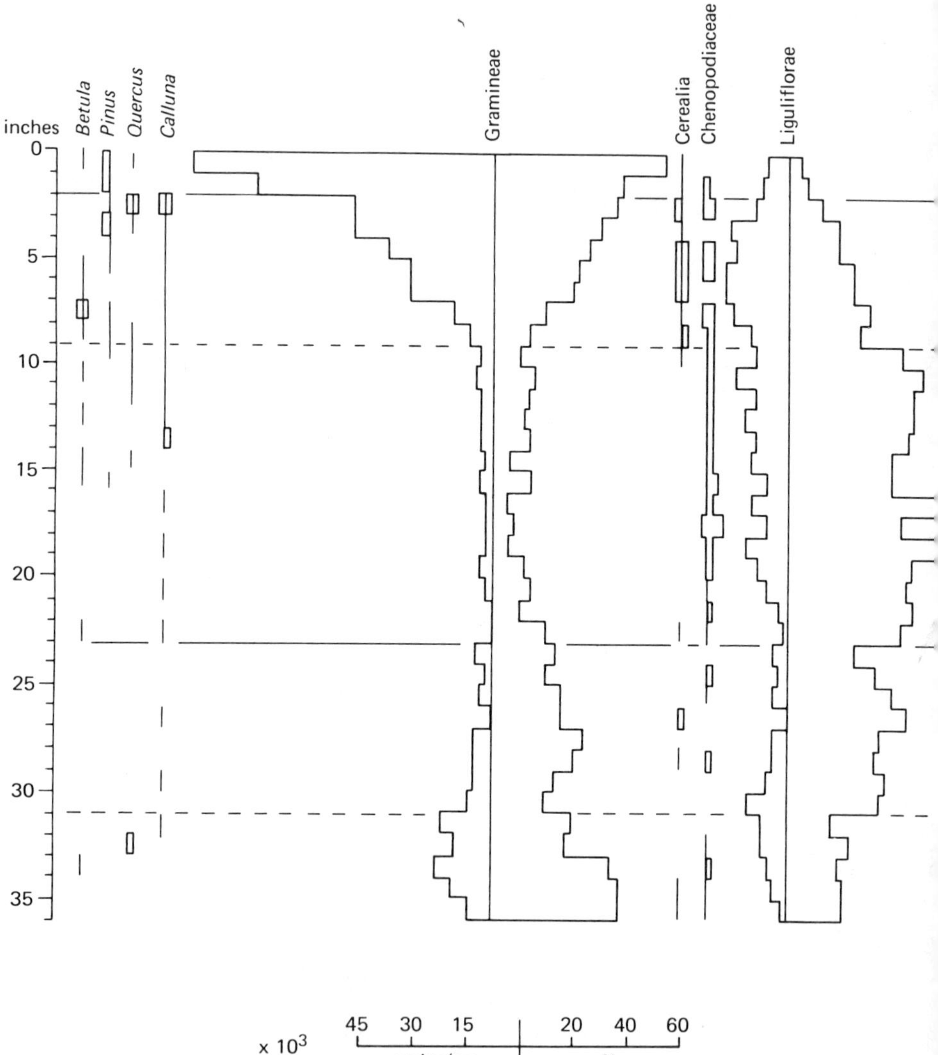

Figure 10. Pollen profile from a cliff face at Halangy Down, St. Mary's, Isles of Scilly.

Where the soil is acid enough to preserve pollen, and yet still permits cultivation, the sequence of soil changes may be clearly demonstrated by the pollen distribution. Such a case was the soil profile at Halangy Down on St. Mary's, Isles of Scilly (Dimbleby *et al.* 1981). A section exposed in the cliff face showed episodes of mass movement of soil, with stable pause phases in

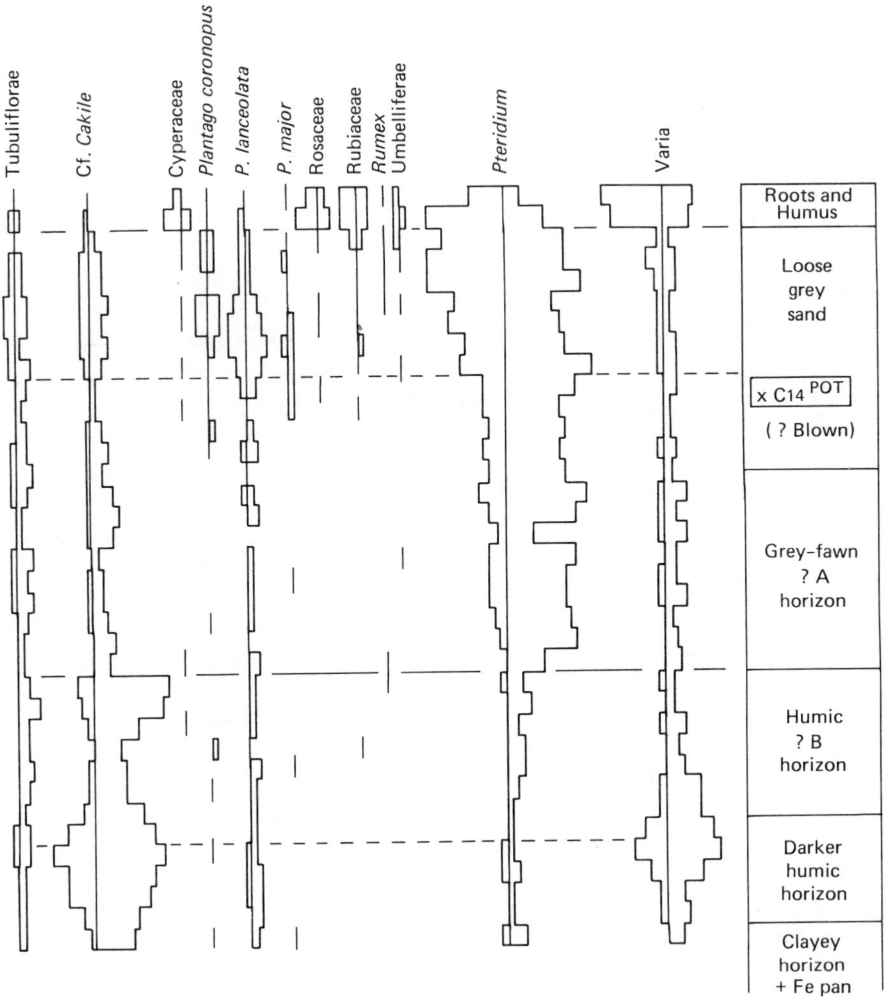

Figure 10. *continued*

between (Fig. 10). It was concluded that the primary soil had been lost, but hillwash had come in, become stable long enough to allow a pollen profile to develop, and that this surface in turn had become covered by a late deposit of hillwash. Here the pollen distribution suggests episodic rather than gradual hillwash. The pollen spectra themselves are a compound of pollen washed in with the colluvium and pollen which has worked down from a surface. There is also reason to believe that there may have been some degree of differential decomposition, leaving a predonderance of resistant types (Liguliflorae,

Pteridium). In such cases, which are not uncommon, the pollen only gives a limited picture of the flora of the site and its environs, but it can give unexpected detail about the sequence of events in colluviation. There is a contrast between those curves which clearly show downwash from a surface (e.g., Gramineae from the present surface; *Cakile* from the surface at 23 inches) and those curves which indicate pollen incorporated in colluvium (e.g., Liguliflorae, 9–23 inches; Gramineae, 23–31 inches). Therefore, in spite of a mixture of pollen of different origins, it may be possible, where there have been pause phases in between episodes of accumulation, to distinguish those pollen taxa which were predominant during the interlude. If, however, the accumulation has been continuous this distinction will not be possible. A similar argument will be used in Chapter 4 in relation to ditch fillings.

It is appropriate under this heading to mention the Middle Pleistocene site of Terra Amata at Nice. Here a complex series of stratified deposits, including actual Acheulean living floors, is buried beneath a multi-aged mass of beach material, dune sand, and several phases of colluvium. Pollen analyses have been made on samples from a large number of these layers; as is usual with Palaeolithic palynology the aim was to establish vegete-climatic differences, and in this case to link with macro remains of plants and vertebrates (Mourer-Chauviré and Renault-Miskovsky 1980). Human coprolites had also been recovered here and a pollen analysis revealed evidence of four distinct plant communities (de Beaulieu 1967).

ARTIFICIAL BURIAL

The burial of an old land surface beneath a man-made structure or an earthwork results in the preservation of a remnant of the past contemporary with the construction. The significance of this had long been recognized in Holland where a correlation between the age of tumuli and the soil type beneath them was apparent; the soils under neolithic tumuli were unbleached, whereas those of the Bronze Age were markedly podzolized. It was realized that the state of the old land surface could be a means of dating barrows. With this end in view, Waterbolk (1950, 1954, 1956) applied pollen analysis to many buried soils, demonstrating a correlation between the development of heath and the type of soil present under the barrows. In this case the stimulus for palynological investigation of the buried soils came from an archaeological interest: the need for dating. But the opportunity presented by the artificial burial of soils was of value to the nonarchaeological sciences too. Shortly afterwards, and long before I became formally associated with archaeology, I was carrying out ecological research (for forestry) into the condition of moorland and heathland soils, and realized that old earthworks, which are often abundant on soils which are today almost unproductive, could hold the key to past conditions. I, too, investigated the buried soils and started to apply pollen analysis to them (Dimbleby 1950, 1955a; Piggott and Dimbleby 1955). This convergence of research by workers in different disciplines, unaware of each other's existence, epitomizes for me the importance of archaeological sites as records of the past, of interest not solely to archaeology. In fact my own ecological research, using such circumstances, was carried on for a decade or more (Dimbleby 1962) before I moved into a more archaeological environment. Having done so, I brought with me my ecological interest, continuing to pursue the pollen analysis of buried soils for other than just dating purposes. Similar developments in the use of pollen analysis have taken place in Holland following on Waterbolk's pioneer work, and have reached fruition in the beautiful work of Groenman-van Waateringe (1978; Casparie and Groenman-van Waateringe 1980). Such has been the value of this approach that now no funerary monument can be excavated in Holland without samples for pollen anaysis being taken from the old land surface and from the mound structure (Groenman-van Waateringe 1979).

This historical note is appropriate here because, as has been shown, the soils buried artificially were the first to attract the attention of pollen analysts, whether within or outside archaeology. The extension of the technique to many other aspects of archaeological sites mentioned in this book generally came later.

The broad principles governing pollen preservation in mineral soils have already been discussed. It was perhaps fortuitous that both Waterbolk and myself started our investigations on earthworks on acid soils. Apart from these principles, however, the question arises as to whether the nature of overlying structure can influence the preservation of pollen. In view of the variability in the preservation of pollen, particularly in less acid soils, this question is difficult to answer, for we have no controls by which to judge. Perhaps it can best be considered in terms of preservation of organic remains in general. The material of which pollen walls are made is one of the most resistant plant products. Other plant remains are rarely preserved in soils, unless they are modified, as by carbonization or mineralization. Occasionally, however, such unmodified plant remains are found, in which case we can deduce that special conditions do obtain. The best example of this that I have come across is Silbury Hill, said to be the largest man-made mound in Europe; it is a flat-topped circular mound standing 40 m high and its base covers an area of 2.2 hectares. Unpublished investigations of the old land surface at the centre of this vast mound yielded recognizable remains of mosses, grasses, seeds, and insects, all uncarbonized, as well as pollen. These will be discussed further below, but the point to be made here is that this preservation occurred on a free-draining terrestrial site. The likely explanation is that the turf core of this mound consumed all the available oxygen through aerobic decay and that once this had been used up diffusion from outside such a huge structure was so slow that anaerobic conditions persisted. In places the buried soil was specked with bright blue vivianite, an iron phosphate characteristic of anaerobic deposits.

It is not possible to say what is the minimum size of structure that will produce such an effect. Where large earthworks have been investigated, as for example the massive banks at Avebury, these have not produced similar preservation, but it has to be borne in mind that whereas Silbury Hill is built on an outlier of acidic Clay-with-Flints overlying chalk, Avebury banks lie directly on chalk. The experimental linear earthworks on Overton Down (chalk) and at Wareham (acid sand) both stood initially over 2 m high, with a spread of 7 m and were demonstrably aerobic at all levels, though they have not yet undergone consolidation over long periods of time (for other details see below). The idea, frequently floated by archaeologists, that a layer of clay over a buried surface will produce anaerobic conditions seems to be very doubtful unless the clay is massive and unbroken.

Besides the possible effect of a large structure on the aeration of a buried surface there is also clearly a potential chemical effect, particularly where the mound itself contrasts strongly in its composition to the underlying soil. This question has often been raised—though not satisfactorily answered—in relation to the effect of chemical leaching in the soil profile; for example, will a buried podzol continue to podzolize after burial? In our present context one might expect that the increased calcification of a noncalcareous buried soil would lead to the decomposition of any pollen originally present. At Silbury, decomposition has clearly not happened but this is understandable if there is a shortage of oxygen. My impression is that even under lesser mounds the effect is not great, if it operates at all. At Kilham long barrow (see below) pollen was present, even though pH values were high. Absence of oxygen may not be the only bacteriostatic effect of burial, but definitive work needs to be done in this speculative area.

In view of the above, what guidance can be given to the archaeologist regarding acceptable conditions for palynology of buried surfaces? First of all, let us consider the nature of the material covering the old surface. The simplest example of this is a slab of stone. Unless such a slab is enormous, one must always be alert to the possibility that lateral movement of water may have brought pollen in from adjacent situations, but some measure of control can be applied by looking at the pollen distribution in the soil, as well as the soil profile itself. If these are both normal and compatible with what one would find in an open site this gives confidence that the slab of stone has indeed preserved the old land surface effectively as it was, and has precluded the contamination of that surface with later pollen. An example of this is a profile examined beneath a kerbstone of a Bronze Age earthwork in Derbyshire, Barbrook II. The soil profile was a well defined stagnopodzol with a thin iron pan, showing no distinction of depth of horizonation as compared with the soils of the adjacent moorlands. The pollen curves, too, showed the characteristic distribution of the pollen with depth, and from this it may be inferred that at the buried surface the pollen spectra do represent the contemporary vegetation. The contrast between this and the present ecologically degraded surroundings is therefore to be seen as an indication of real changes that have taken place since the Bronze Age.

A horizontal megalith at a chambered cairn, Bedd-yr-Afanc, in Pembrokeshire was investigated for dating purposes. Under the stone was a podzol profile (Fig. 11), the top three inches of which proved to be very rich in pollen. As the spectra from the three samples were reasonably consistent among themselves this zone was treated as a unit. The analyses are shown, based on total tree pollen (AP) in the upper part of Figure 12. A search was made for a nearby peat deposit for cross-referencing, but the best that could be found was a shallow deposit about half a mile away (Hafod Tydfil). Samples from this were analysed and produced the lower diagram in Figure 12. Both sets of histograms are plotted at the same scales and on corresponding axes, so vertical comparison is simple. It is apparent that the Bedd-yr-Afanc soil pollen can only be correlated with that part of the peat profile above 55 cm. This appears to be late, about the Subboreal (VIIb)/Subatlantic (VIII) transition. However, this peat profile is not ideal, so comparison was also made with three published profiles from west and southwest Wales. One was Borth Bog (see Godwin 1975:Figure 30), and two were from Tregaron Bog, profiles W. 27 and S.E. 10 (Godwin and Mitchell 1938). In each of these cases the only good match with the soil pollen was at a level interpreted by Godwin as being near the VIIb/VIII transition. It seems indisputable that the old land surface (OLS) buried beneath the stone was of this date. Whether this truly dates the monument depends, of course, on whether the stone was in a primary position.

On Bodmin Moor, Cornwall, Clews has similarly selected sampling points beneath large flat stones in order to eliminate any possible effects of downwash (Brisbane and Clews 1979). The contrast between the pollen sequence in his buried soil and the spectrum from the modern land surface indicates that his assumption is justified.

It is difficult to specify how large a stone has to be to provide protection, as we have no data on the point. The effect of any large relatively impermeable object in a porous matrix will be to intensify percolation around its edges; as long as the water moves vertically downwards no problem arises, but there is the possibility of lateral movement. However, a large slab may give rise to such soil compaction that water is unlikely to flow underneath it. This is pure speculation, but the possibilities of error must increase the smaller the object.

Figure 11. Soil profile beneath a megalith, Bedd-yr-Afanc, Pembrokeshire.

Heim (1979) refers to sampling underneath stones and potsherds, though it is not clear from his account what results he obtained. Following a similar reasoning I was asked to carry out pollen analyses on spot samples taken from immediately beneath eight stones on a Dutch mesolithic site, Bergumermeer; the stones were all believed to have been set in their positions contemporaneously with the mesolithic occupation. Good counts were obtained in all cases, but there was so much variation in the spectra between the different sampling points, particularly as regards the non-tree pollen, that no meaningful interpretation could be made.

Also on a mesolithic site, in Belgium, Vermeersch *et al.* (1974) similarly tried to date the culture by pollen analysis of samples from immediately beneath stones believed to be *in situ,* such as hearth stones. Though the stones were described as large, extrapolation from the site plan indicates that they were only about 20 cm in diameter. Three stones were investigated, one of which gave an unacceptable Subboreal date and the other two, though giving similar spectra, gave a date younger than was expected. A possible explanation given for this was that contamination with younger pollen may have taken place via root channels. It was observed that fine rootlets formed a felt mat beneath the stones. It was mentioned above that water may be channelled around stones, and as fine roots seek out water such root activity is to be expected. Because of such marginal effects there may be a minimum size of solid object necessary to protect the underlying surface; at present we can only guess what that size is. One final point strikes me; if hearth stones are selected as being *in situ,* one would expect any contemporary pollen to have been destroyed by heat, in which case any pollen found today should be intrusive. It is clear that much critical observation needs to be made before this approach can be accepted as reliable.

Where the old land surface is covered by a porous deposit, particularly if that deposit itself contains pollen, there may also be uncertainty about contamination. How far is this uncertainty justified? It was mentioned earlier that in soils pollen movement downwards is slow; a rate of 10 cm in 300 years was suggested as typical. However, as we have seen, not all pollen of the same taxon and of the same age travels at the same rate, giving distribution curves of the type shown in Figure 3. There may therefore be a few grains which reach considerably deeper than the main mass, but these may be at such a low concentration that they do not significantly affect the lower layers. This may particularly be the case with a buried surface. As demonstrated earlier, soil surfaces are much richer in pollen than the lower horizons. Consequently a few grains reaching a buried surface from above may be of no consequence.

But are we justified in quoting rates of movements based on soils *in situ* and applying them to, say, earthworks, where the body of the earthwork is made up of soil which has been disturbed, loosened, and redistributed? The only

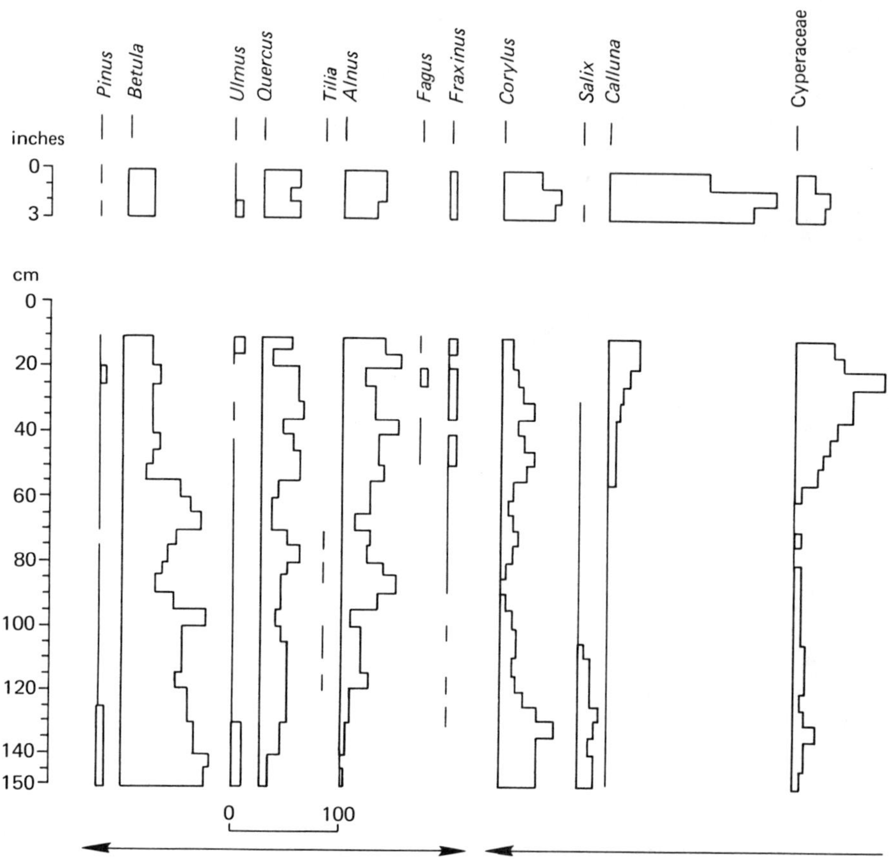

Figure 12. Pollen diagrams from Pembrokeshire. Upper diagram: pollen from the surface of the buried soil of Bedd-yr-Afanc, shown in Figure 11. Lower diagram: pollen profile from a small bog at Hafod Tydfil.

way to judge this is to look at some examples. Several analyses were published (Dimbleby 1962a) showing the distribution of pollen from the present surface of round barrows and examination of these suggests that the depth of penetration by post-barrow pollen ranged from 4 to 13 inches (10–33 cm; Dimbleby 1962a:Figures 2, 3, 4, 8, 9). It is also noticeable in some of these examples that below these levels any stratification, for example due to turf structure or artificial layering, is not blurred but is clear and distinct, suggesting that the pollen is immobilized below the level of active soil processes. On this evidence, therefore, it appears that a cover of 16 inches (40 cm) should give adequate protection against contamination from an exposed surface above. That this is so is shown by a series of anlayses carried out through low banks

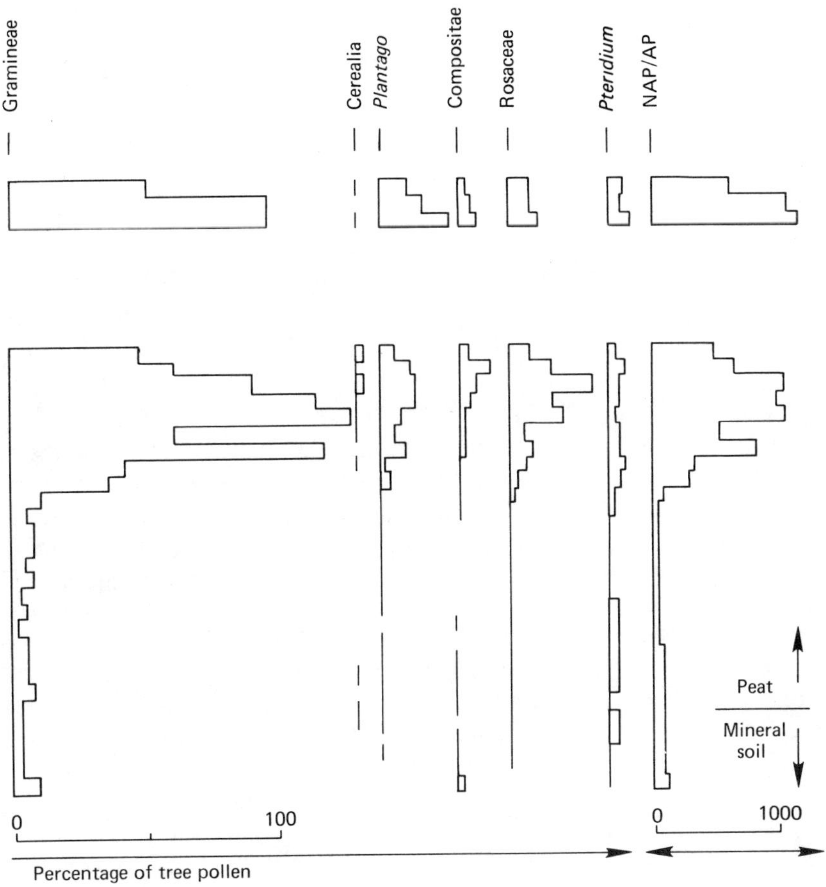

Figure 12. *continued*

in the New Forest (Tubbs and Dimbleby 1965). Because of their low stature these banks had gone unrecognized, and nothing was known of their age or purpose. There had certainly been wastage of the banks, since their associated ditches were silted up, but those examined were covered with vegetation. Two of them (Hincheslea and Holmsley) were sampled from their present surface down to and through the old land surface they covered. Though the age of the banks proved to be widely different, in neither case did the post-bank pollen, recognizable by the characteristic V-shaped distribution curve, interfere with the interpretation of the pollen in the buried soil. The circumstances here are, of course, marginal and one would prefer a greater depth of overburden, but they confirm the conclusions reached above.

More recently a similar conclusion has been reached in regard to the 'reaves'—ranch boundaries of Bronze Age date—on Shaugh Moor, Dartmoor (K. Smith *et al.*1981; Balaam *et al.* 1982). These linear earthworks, extending for thousands of metres, were constructed as a low (6–10 cm) earth bank thrown up from a parallel ditch, and then capped first by a wooden fence, subsequently replaced by a stone wall. Walls of this type were typically 1.5 m wide and 30–50 cm high, and consisted of stone facing to a dumped stone core. Such a structure hardly seems a likely barrier to pollen down-wash, yet neither the buried soil under the bank nor the bank surface, only protected by the tumbled-down wall, was recognizably influenced by modern pollen.

DATING OF THE BURIED SOIL

At the beginning of the section on artifically buried soils reference was made to the early work of Waterbolk, in which he was able to put barrows into their cultural period by the pollen spectra from the buried land surfaces. Pollen spectra analysis can never give precise dating, of course, and it has today largely been superseded by radiocarbon dating, but it is still useful on occasion. For effective dating there must be recognizable differences in pollen spectra from different periods. A widely used distinction is that between layers that do not have agricultural pollen in them and those that do. This should separate pre-neolithic levels from later ones. However, as we learn more about the origins and spread of agriculture even a distinction as fundamental as this cannot always be relied upon. Heim (1979) has reported significant quantities of cereal pollen beneath a neolithic tomb in Belgium; surprisingly the tree pollen indicated an Atlantic age for the deposit. Any doubt that the origin of the cereal pollen is removed by the fact that carbonized cereal grain was also present. Recently Edwards and Hirons (1984) have discussed the evidence for cereal pollen before the elm decline in Britain and Ireland. Most of the evidence comes from nonarchaeological deposits, but at the site of Ballynagilly in Northern Ireland, discussed in other contexts at the end of this chapter, a single pollen grain of cereal type was found at a level dated by radiocarbon at about 3800 b.c. (see Ch. 7 for further discussion of cereal pollen).

In Holland, as elsewhere in Western Europe (but not Britain) the establishment of beech (*Fagus*) as a forest component provides a time marker, for it nicely separates the neolithic period, in which it is not present, from the Bronze Age, when it is consistently represented in the pollen spectra. A very clear example of this is given by van Zeist (1963) in which he was able to distinguish the early phase of the tumulus Te Mol as neolithic, whereas its enlargement took place in the Bronze Age. Tree pollen correlations of this sort are the most reliable means of dating in absolute terms; they are the basis on which the postglacial forest sequence is recognized. Another example is the elm decline, mentioned above, which marked the onset of the Neolithic and which we shall have cause to refer to again later in connection with buried mesolithic sites. However, it must be recognized that not all changes in the tree and shrub pollen are climatically determined. We know that the disappearance of *Tilia* (lime)—and probably elm too—was the result of man's impact on the forest (Turner 1962; Iversen 1958). It has been suggested that the great expansion of hazel (*Corylus*) in the Boreal period may have been an anthropogenic effect of the strategic use of fire by mesolithic man. Other trees which are certainly affected by human pressures are birch (*Betula*) and ash (*Fraxinus*), and most of the shrubs. In Britain beech expanded somewhat later than on the Continent, though we know that pockets of it had been in southern Britain since the Atlantic period. We still do not know why it took off, but it seems likely that man triggered the change in some way. In the New Forest, where beech is a conspicuous component of the forest canopy, it appears that it only became important in the last two or three centuries (Tubbs 1968) though undoubtedly locally present much earlier.

If man himself has affected the distribution of trees in this way it means that absolute dating by this method can only be relied upon where there is no reason to suspect anthropogenic influence. On the other hand, *relative* dating may still be possible. If, for instance, the effect of man's use of the land was progressively to eliminate some species or favour others, these very changes may form a recognizable sequence, from which relative ages may be inferred. Such a chronosequence will be more reliable in an area of ecological uniformity and limited size (Dimbleby 1955b). As we have seen, it was a similar sequence of ecological events, particularly the spread of heather (*Calluna*) as shown in the pollen analyses of the Dutch barrows, that Waterbolk (1954) used to distinguish neolithic from Bronze Age mounds.

A similar approach was used to demonstrate that heather moorland in northeast England was established relatively late in the sequence of ecological changes that began with forest clearance in late neolithic times (Hayes 1967). Here monuments of different ages occurred side by side on the same stretch of moorland, so meeting the criteria specified above.

BURIED SOIL PROFILES

It is only appropriate here to include a consideration of the soils buried beneath monuments insofar as they hold features of interest in the application of palynology. We shall be looking later on at the deductions that may be made on the basis of pollen analysis about the structure of monuments; here that aspect will be excluded, and we shall concentrate on the soils themselves. This can most effectively be done by examining a few examples of the pollen evidence from soils buried beneath earthworks.

One of the most striking examples of the value of such evidence was the soil buried beneath the rampart of the hill-fort called Caesar's Camp at Keston (Dimbleby 1962a). The buried soil was a strongly developed podzol, but the pollen sequence showed that it had been associated with oak forest. There was no indication from the depth distribution of the pollen that there had been any other type of vegetation cover; the only trend towards more open conditions was at the top of the buried profile, where there was an increase of the light-demanding birch (*Betula*) and bracken (*Pteridium*), though the oak remained strongly predominant. Whilst the ecological interest of this case is considerable, the archaeological significance should not be overlooked. Although it must be admitted that only one sampling point was chosen, it raises the question as to why a hill-fort should be constructed in woodland. I know of at least one other parallel, at Cogges, near Witney, Oxon.

A different problem was posed by the pollen analysis of the very shallow soil beneath the banks of the causewayed enclosure of Windmill Hill. Here there was a conflict in the interpretations of the evidence from molluscs, which suggested shaded conditions, and the pollen of the buried soil surface, which clearly indicated open country with agriculture. The likely explanation is that a deeper soil had been truncated, probably artificially; this would mean that the molluscs found there would be those normally found in the lower horizons of such soils, that is, shade-tolerant species. However, a truncated profile would have removed the more pollen-rich upper layers; the fact that there was pollen present suggests that the truncated surface has been exposed long enough for pollen to collect a second time. This pollen was strongly agricultural, and indeed the soil truncation may have been the result of an agricultural practice such as paring and burning. This example and others from calcareous soils (Dimbleby and Evans 1974) underline the importance of recognizing the significance of different types of biological evidence. To the archaeologist looking for an interpretation he can use, one form of biological evidence seems as good as another. But pollen grains are very different from snail shells; they get into the soil in different ways, they survive (in calcareous soils) for different lengths of time, and probably indicate ecological conditions over different spatial dimensions.

Another calcareous buried soil, already mentioned, was that under the long barrow at Kilham, East Yorkshire (Manby 1976). The top 20 cm of the buried soil were sampled serially for pollen, which was present in countable quantities only in the top 12 cm of the soil and in the lowest sample (18–20 cm). The soil itself was originally decalcified, being of the *sol lessivé* type, and contained no mollusc shells. The pH values in the horizons sampled ranged from 7.5 to 8.0, no doubt due to percolation from the overlying chalk mound. It is instructive that despite the high pH values pollen was present, suggesting that pH alone should not be taken as a measure of microbiological activity that would result in the destruction of the pollen. At the point of sampling there was a depth of only 15 cm of calcareous ploughsoil covering the palaeosol, so the latter would certainly have been fully aerated. The pollen profile clearly fell into two sections, the lower (a) being separated from the upper (c) by a zone virtually devoid of pollen. Interestingly the lower section indicated much greater treelessness than the upper, suggesting a period of agriculture followed by a phase with woodland regeneration, with lime (*Tilia*) as the most abundant tree species. The details of the profile are complex pedologically and palynologically and need not concern us here, but this site underlies the fact that interpretation may suggest a much more complex cultural history than the archaeologist usually allows for, one that needs more than one line of evidence for its reading.

Some divergence between archaeological expectation and the palynological evidence occurred at a site of very different age—the Anglo-Saxon ship burial at Sutton Hoo in Suffolk (Bruce-Mitford 1975). A number of pollen profiles were studied from this site, but here we will consider only one, that from the soil beneath the ship-mound itself. Earlier environmental work (not using pollen analysis) had suggested that the cemetery had been constructed on old heathland, but the pollen analysis did not support this. Although the soil was acid and pollen well preserved, the pollen record showed abundant agricultural pollen and the merest traces of heather (*Calluna*) pollen and bracken (*Pteridium*) spores (Fig. 13). The profile can be divided into three zones, the lowest (Z) possibly representing a soil that was buried long before the cemetery was built. The other two zones were interpreted as one profile, the upper part of which (X) had been ploughed, whilst the lower part (Y) showed stratification that had obviously not been disturbed by the plough. The soil here was too acid to permit vigorous earthworm mixing, so cultivation is the only alternative explanation. Zone Y, below the plough zone, suggests a change from pastoral farming, with high values of ribwort plantain (*Plantago lanceolata*) in the lower layers, to arable farming, indicated by plants more characteristic of disturbed soil (e.g., *Rumex acetosella*). At the top of zone Y and throughout zone X cereal pollen reached values unprecedented in my experience. A glance at the curves in this zone will show that both the frequencies and the percentages of each taxon are virtually constant throughout the zone, a sure sign of physical mixing of the soil.

In the example just quoted it was pointed out that the soil at Sutton Hoo was too acid to permit earthworm activity. A specific study of what happens when earthworms are or are not present was incorporated into the programme for two experimental earthworks, both of them a bank and ditch, the one built

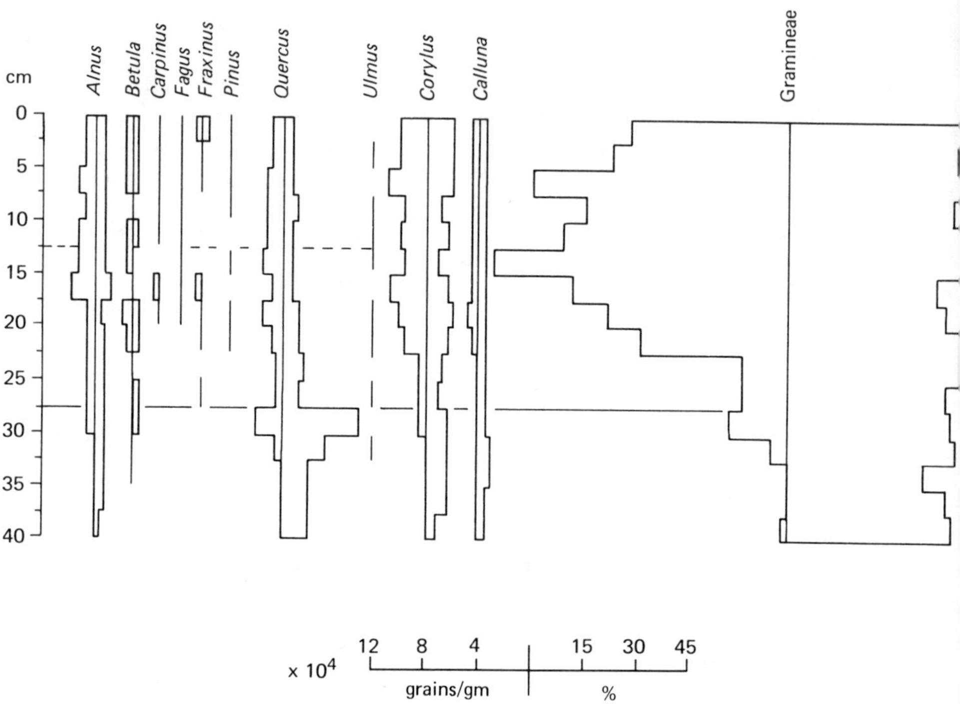

Figure 13. Pollen diagram of a buried soil beneath Ship Barrow, Sutton Hoo, Suffolk.

on a chalk soil at Overton Down, Wiltshire (Jewell 1963) and the other on acid sand at Morden Bog, Wareham, Dorset (Evans and Limbrey 1974). In both cases the ground surface was covered with spores of *Lycopodium* immediately prior to the building of the earthworks. These spores were chosen as being comparable in size with pollen, easy to identify, alien to the local flora, and resistant to microbiological decay. When the earthworks were sectioned at given time intervals, three series of samples, each running from the bank, through the old land surface and down into the subsoil, were taken at each sectioning. Comparative results over a 4-inch range of depth are shown in Figure 14. At the Wareham site, in which no earthworms are present, the spores have all remained within an inch of the level at which they were deposited for a period of 9 years. At Overton Down, on the other hand, in a period of only 4 years they could be found not only as deep as 8 inches into the soil, but also in considerable quantities well up into the bank itself. Visual

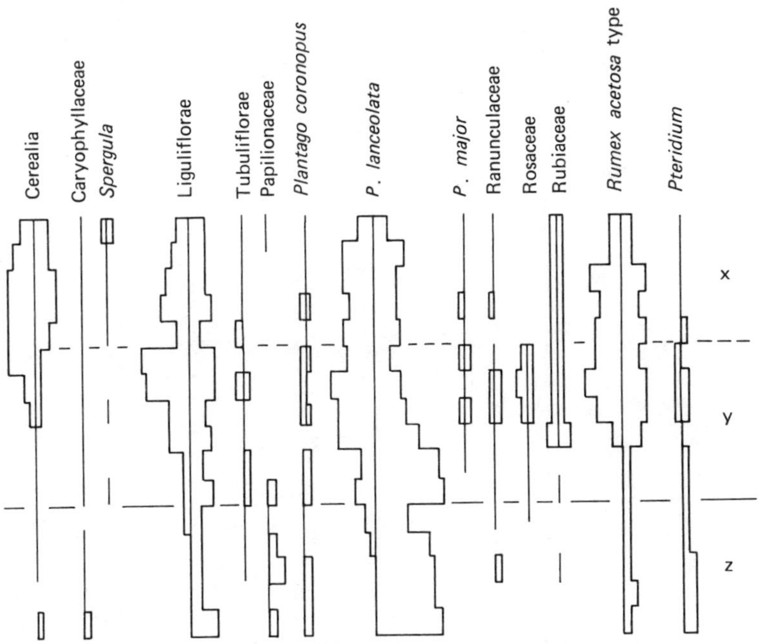

Figure 13. *continued*

observation of the sections showed earthworms to be present in the buried soil and to be working up into the bank and casting in the interstitial spaces. Over a period of time the spores will be distributed throughout the layers in which worms are active, producing the effects noted in Chapter 1.

At many points in this book reference is made to the pollen evidence of major vegetation change brought about by prehistoric man, particularly on acid soils (see discussion in Ch. 1). A very informative example of this, in which the pollen analysis was combined with micromorphological studies on the buried soils, was the work done on the Dartmoor reaves, referred to above (Balaam *et al.* 1982). Not only did the soil pollen indicate changes of vegetation but it was shown that the buried soils themselves had already undergone severe deterioration even by the Bronze Age, doubtless as a result of primary clearance and land use. It was also demonstrated that the effect of constructing the initial bank of the reave was to stop further pedogenesis (even though lateral water movement continued to occur in the soils), and, perhaps more surprisingly, the construction of the stone reave itself halted pedogenesis on the surface of the bank. In some cases it was found that the buried soils were truncated prior to the construction of the primary bank. Whilst this no doubt occurs in connection with land use operations, as suggested at Windmill Hill,

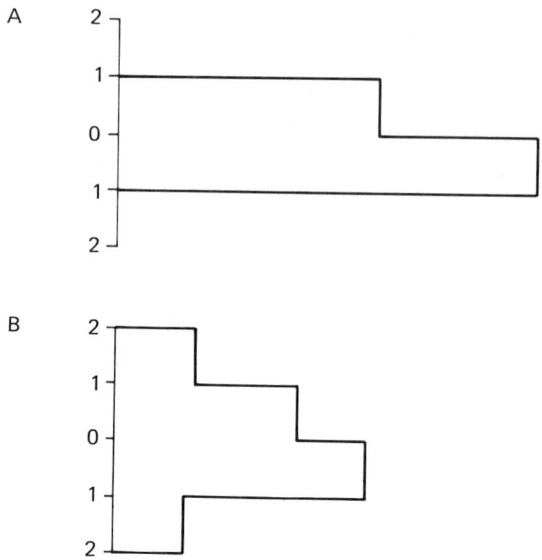

Figure 14. Movement of *Lycopodium* spores in experimental earthwork A, Morden Bog earthwork after 9 years; B, Overton Down earthwork after 4 years. Vertical axis: depth in inches. Horizontal axis: sum of percentages from 3 profiles at each site.

it certainly occurs commonly in connection with the construction of earthworks. Here, too, the pollen distribution can shed light on the process, but consideration of this particular aspect will be deferred until Chapter 4 when we shall be describing the value of pollen analysis in elucidating the structure of monuments.

CORRELATION BETWEEN ON- AND OFF-SITE SPECTRA

It is commonly held that the most reliable sort of evidence of the contemporary ecological conditions of a site can only be obtained from an adjacent peat or lake deposit, if there is one available. Frequently, of course, such a deposit is not available, and one therefore has to rely on the old land surfaces buried on a site. How comparable is evidence from this source to that obtained from a peat deposit? Groenman-van Waateringe (1979) has put it this way: In order to understand the impact of a particular settlement on the surrounding environment it is not the regional vegetation, but the tree and plant growth in the immediate vicinity of the settelement which must be examined. Usually a nearby pool, river or creek, perhaps now totally overgrown or silted up, is sufficient for the purpose. It is especially important to take samples from as near to the settlement as possible if one is especially interested in the weed vegetations (weeds of meadows and arable land and of the crops cultivated on the fields). She goes on to present detailed analyses from old land surfaces and elsewhere in a series of barrows, and she compares these with the relevant parts of the nearest peat profiles, some of which were several kilometers distant. Pollen analyses from old land surfaces, one would expect, being in direct association with the airborne pollen of the site itself, should give the most precise representation of the local pollen rain. However, there are major sources of bias, which will be discussed in Chapter 7. As we have seen, pollen data from this context have been regularly used for archaeological environmental studies in Holland, Britain, and a number of other countries. Elsewhere, however, such work has relied upon pollen data from peat bogs and other waterlogged deposits, and it must be apparent that such data cannot be related with precision to particular sites; there must always be an element from the wider area. There has been very little work done in establishing correlations between these two potential sources. However, the work of Pilcher in Northern Ireland has provided us with some facts on the subject. At Ballynagilly, County Tyrone (Pilcher and Smith 1967), an early neolithic site, a full pollen profile was taken from a nearby valley bog, and mineral soil profiles in the vicinity of the site were also sampled. Evidence was found not only of forest clearance but also of soil erosion. The blanket peat did not begin to form until after the Bronze Age. Here no samples were apparently taken from the settlement site itself, but at another neolithic site, Beaghmore, also in County Tyrone, not only was nearby basin peat sampled, but so was the old land surface under a cairn and the fill of a ditch (Pilcher 1969). There were major differences in the proportions of some of the weed pollens; *Plantago,* for instance was only 5% of the total at the appropriate level in the peat, but over 20% in the bottom of the ditch. Similar high levels of *Plantago* were recorded in the buried soil at Goodland, described above (pp. 49–50). Pilcher concludes that at Beaghmore, as the two sampling points were only 60 m apart, the soil pollen must represent 'parochial' conditions.

Caseldine (1983a) stresses the corollary to this: that until we know more about areal relationships in pollen deposition, pollen analyses of a buried soil cannot be regarded as representative of wider spatial conditions.

In the palaeoenvironmental report on Shaugh Moor, Dartmoor (Balaam *et al.* 1982), it was similarly concluded that the pollen from the soils buried beneath the reaves largely represented local pollen; indeed, by studying the buried profile at intervals along a stretch of 350 m of the reave it was possible to demonstrate vegetation zonation related to altitude—in other words to produce an ecological transect of a Bronze Age landscape. At the same time it was found possible to link the chronological changes of vegetation with a standard peat pollen profile from a bog some 500 m distant. This showed clearly that some of the pollen in the bog sequence, notably *Alnus,* was local to the wet areas surrounding the bog, as it was much less prominent at any level in the soil profiles. Balaam says that whilst the pollen from the buried soils did not add anything new to the general picture obtained from the peat profile, it did allow the spatial study of vegetation in the period before reaves were built, and moreover it made it possible to correlate the peat pollen zones with the archaeological material with much greater accuracy.

How near to a site does a peat deposit have to be before it truly reflects the conditions of the site itself? Welten (1967) studied a transect of profiles from a lakeside settlement outwards and found that the proportions of certain pollen types changed dramatically over the first few metres from the settlement edge (see further discussion in Ch. 7). The old land surface under a funerary mound may or may not be comparable with the surface of a site on which people were living and it can be shown that the proportions of some pollen types can be biased in the latter situation (see Ch. 7). It does seem, however, that at least as far as cultural pollen goes, spectra from adjacent peat deposits may not even approximate those from buried land surfaces on a site. The assumption that there is such an approximation is inherent in all the examples referred to above and in many others in the archaeological literature, but clearly much more critical exploratory work needs to be done before these underlying assumptions can be accepted as valid. How much more doubtful, therefore, must be environmental conclusions based on waterlogged deposits situated at much greater distances from the sites in question. If it is only approximate dating that is required this may not be a problem (cf. Bedd-yr-Afanc, p.57), but the interpretation of cultural activity is beset with difficulty.

The imponderability of relationships of this kind is underlined by two recent papers on work in Finland. Vuorela (1982) carried out pollen analyses on a soil at a Corded Ware site and compared them with analyses from an adjacent peat deposit. The soil analyses provided greater detail than the peat analyses, but in broad terms they agreed in giving a picture of a nonagricultural fishing and hunting economy. Such parallelism, however, was not found in a similar investigation (Vuorela and Aalto 1982) of a neolithic dwelling site. Although a peat deposit was close to a known Stone Age settlement, the pollen spectra from it showed little recognizable evidence of human interference. However, at one level in the peat there was a massive accumulation of macroscopic remains of the water chestnut, *Trapa natans;* there were many endocarps (nuts) which had been cracked open, probably by Stone Age man. Were it not for this material, and the preknowledge of a Stone Age settlement in the vicinity, the presence of a dwelling site in the neighbourhood would not have been suspected. Vuorela and Aalto conclude that 'it is not the distance from the dwelling site that is decisive for the evidence of settlement in pollen diagrams, but the nature of the settlement'. Further discussion of pollen analysis of settlements is given in Chapter 7.

4
INTERPRETATION OF STRUCTURES FROM POLLEN ANALYSIS

INTRODUCTION

Arising from the usual distribution pattern of pollen in a soil profile, where the overall frequency drops steeply from the surface downwards, we have seen that it is possible to recognize old surfaces beneath earthworks and to some extent to read the vegetation history prior to burial. On the same basis it is also possible to learn something about the structure of earthworks and sometimes about changes, such as enlargement, that have been made in them.

Earthworks, by definition, are made of earth. That term, however, is not explicit in identifying the source of that earth. There are various possibilities: earth may be scraped up from the surface of the surrounding area; it may be dug up, as for instance, in the excavation of a ditch; or it may be transported from some more distant area; or in any one structure it may have been provided by a combination of any of these.

It must be admitted that it is often easier to extract botanical information than to interpret it usefully. For instance, I have found pollen in low but countable quantities in mud-brick from Çatal Hüyük; it was mainly of the family Chenopodiaceae, reminiscent of the pollen from a wadi in Jordan (p. 36) and open to a variety of interpretations. However, the presence of other plant remains has also been recorded, notably in adobe used by North American Indians, and further investigation on these lines might prove profitable. Another similar material is clay or daub used, for example, in wattle and daub construction. Here too plant remains have been found, mostly seeds of weeds and crops (Arthur 1965), but I am not aware of any pollen analyses on such material.

TURF STRUCTURES

The word *turf* as used here, and as generally used in archaeological description, implies a cut piece of the land surface, including the surface humus (perhaps with vegetation), the rooting zone, and the mineral soil beneath. It may not be of uniform thickness, depending on the cutting implement used, but at its greatest thickness it may be 15 cm or so thick. We are not talking only about purely organic turf, as from a peat bog, though this would not be excluded, nor are we talking just about grass sward. Turves can come from grassland, heathland, farmland, or even from a woodland floor.

Turf structure is often recognized visually (Dimbleby 1962a:Plate III, 3 and 4; Plate V, 1; Drewett 1975:Plate II; Ashbee and Dimbleby 1976:Plates 1 and 2). Where the turves are cut from heathland the mineral soil attached to them, being the A horizon of a podzol, is often strongly bleached, forming a strong colour contrast with the dark humus layer. When stacked the effect is unmistakable. In such cases it is usually obvious whether the turves are in their normal attitude or inverted, though not always. However, turves may be cut from a soil surface which does not show the same contrast. Even on podzolized soils this may occur where there is deep humic staining of the A horizon. On soils which are not bleached, however, piled turves may be much less easy to recognize. At Sutton Hoo, for example, observers differed in their opinions as to whether the ship barrow was constructed of turves or not (Bruce-Mitford 1975); there was no well-defined turf line at the surface (I have already given reasons for believing the land had been cultivated, see pp. 65–66 above) and the sand was of a uniform colour. I have no doubt that the question could have been resolved by pollen analysis, but as far as I know no samples were taken for this purpose. At the other extreme of soil types, earthworks built on rendzinas have also been recognized as having turf structure (i.e., Snail Down, Thomas and Thomas 1955); the humus layers are characteristically strongly developed and contrast with white mineral soil beneath, but over time the humus tends to decompose and become impregnated with lime so that the contrast is reduced. Attempts to apply pollen analysis to such a turf structure might be unprofitable because of the poor pollen preservation and earthworm activity in such soils.

Pollen analysis through turf structures of earthworks on acid soils can, however, shed light not only on the structure itself but on the origin of the turves. It is common for structures on acid soils to have a well-defined turf core which is then covered with soil material without visible structure. This would make sense if an area of ground was marked out as a source of earth for the structure. The turves would be cut and stacked centrally in the new earthwork, and then material from beneath the turf would be put on top of the core. However, Ashbee has calculated, for a round barrow at Moor Green in Hampshire, that the amount of turf used for the core was far greater than could have been provided by stripping only the area of the ditch (Ashbee and Dimbleby 1976). In this case pollen analyses were carried out on a number of turves and the results suggested that the turves had been cut from different parts of an area of woodland probably cleared by fire. The analyses indicated that some turves had come from the margin of such an area where it adjoined forest, whilst others showed less influence of tree pollen, consonant with a position more central in the opened area. Pollen analysis offers scope for investigation of this sort of problem, though in this case it was not pursued further. Incidentally this may be seen as another example of the unexpected environment mentioned in the last chapter. There the case was of a hill-fort apparently set in woodland. This round barrow at Moor Green was set in open woodland on land which had never been cultivated, despite the fact that the reason for excavating the barrow was that it took up a sizeable area in a present day farmyard and the farmer wanted to get rid of it. Pollen analysis may not always give the answer the archaeologist expects—or hopes for.

As mentioned above, it is often possible to tell whether turves are all stacked the same way up, though this is not always clear visually (Fig. 15). There is usually no difficulty of telling from the pollen analysis. The surface of a turf has a high pollen frequency (one of those at Moor Green had a frequency of over 7 million grains per gram of soil, one of the highest values I have ever come across); if this turf is upright and is covered by another upright turf, there will be a sharp change in frequency at the interface; in contrast, in each turf the frequency will fall gradually from the surface down

Figure 15. Turf stack in round barrow, Minsted, Sussex.

through the soil. An inverted turf will show this gradual change in frequency going upwards through the turf.

Some examples will make this clear. In a barrow at Minsted, Sussex, a short pollen sequence (see Fig. 27) was taken from the subsoil, through the old land surface, into the base of the mound (Drewett 1975). For present purposes it is irrelevant that there is a buried soil of mesolithic age beneath the barrow—this will be returned to later—but it is apparent that the pollen curves above the Bronze Age land surface are a rough mirror image of those in the buried soil itself, even to the high representation of *Hedera* (ivy) pollen. Clearly there is an inverted turf lying directly on to the Bronze Age land surface, and the close parallel in the pollen spectra must mean that it was cut from very nearby.

In such analyses it is the commonest case that turves were laid surface downwards, though in the Minsted barrow described above they were mostly upright. Examples of inverted turves which are clearly seen at Wilsey Down, Cornwall (Dimbleby 1963a:Figure 1), the Burton Howes, North Yorkshire and Winter Hill, Lancashire (Dimbleby 1962a:Figures 8, 9, 31). The section of the barrow at Winter Hill also shows one turf that has been placed in the upright attitude.

It was mentioned above that pollen analysis of turves could give an indication of the contemporary vegetation at the site where the turf was cut. For this purpose the turf must be treated exactly like the surface of a soil, which of course it was. It is therefore important that the sample from which the ecologically significant pollen spectrum is obtained should be as near as possible the actual surface of the turf. Reference to the diagrams quoted above will show that the pollen spectra of contiguous samples through a turf, as through a soil, can vary considerably. On the other hand, it is not always easy to obtain a sample of the very surface of a turf, particularly if it is lying on another turf surface. It also has to be borne in mind that the organic layers of a turf, such as the raw humus of a heath soil, or even the spongy-structured mull layer of a grassland soil, will be liable to considerable compression under the weight of the overburden. This was clearly shown in the experimental earthworks on Overton Down and at Wareham; the compression virtually all took place in the turf core during first two years of burial under the earth banks, 1.8 and 1.5 m high respectively. Such compression must make it even more difficult to limit sampling to the actual surface of a turf, without inclusion of adjacent layers which would have different pollen spectra. I must confess that my early sampling at 1-inch (2.5 cm) intervals was probably not close enough to give reliable ecological evidence. This factor needs more precise investigation: Whittington (1983) has recently been carrying out close sampling on old land surfaces buried beneath blanket peat, but the circumstances are not exactly parallel to buried turves, not least in that the latter are usually freely drained.

The change in pollen content as one goes down through a soil or turf raises the question of the completeness of the buried soil itself. In discussing the soil buried beneath the causewayed enclosure of Windmill Hill it was suggested that the shallow soil profile there did not represent its natural dimension, and evidence suggested that it was truncated. It was also mentioned above that a barrow on the chalk at Snail Down was apparently composed of turves. In fact there were two barrows here, both built of turves, but the soil beneath one of them was significantly shallower than the other. The excavators (Thomas and Thomas 1955) deduced that one barrow was built of turves cut from the adjacent area, and the second was sited on ground which had been deturfed.

Certainly it is a common situation to find truncated soils beneath earthworks. That soils are tuncated may often be inferred from the absence of a visible turf line, though even the presence of one may not be proof that the profile is intact. Under the Winter Hill barrow referred to above there was an apparent turf line, but the pollen analysis shows that there is a sharp break between this line and the underlying soil; clearly the soil has been truncated and has been exposed long enough for a new turf line to develop. Comparison of the pollen analysis of this feature and that of the turves that make up the mound shows that the latter are markedly richer in tree pollen, and that the proportions of the woody species are not the same as for the turf line. The interpretation of these differences was that the barrow was built in a treeless clearing whereas the turves came from woodland or glades.

There are other cases, however, where there has been no re-establishment of vegetation on the truncated soil and where the inverted turf, if reversed, would fit upon the truncated profile both pedologically and palynologically. An example of this is the barrow Burton Howes 1A, referred to above. This was one of a group of barrows, and, as at Snail Down, may have been built on land recently deturfed for the construction of one of the others. Two of the group were examined palynologically; in each case the turves of the mound were the only possible sources of contemporary pollen since no turf line existed on the buried soil. The matching of the turves with the truncated soil suggested that they were derived from the immediate locality and not, as at Winter Hill, brought from a distance. It was shown that the tree pollen spectrum in each case was virtually identical, but one had a lower non-tree pollen percentage, suggesting that it may have been the later in date of construction, or from a different point in a clearing.

SUBSOIL VERSUS TOPSOIL CONSTRUCTION

It would be rash of me to venture into a discussion on the fundamentals of earthwork construction. Was a ditch dug because a ditch was wanted, or

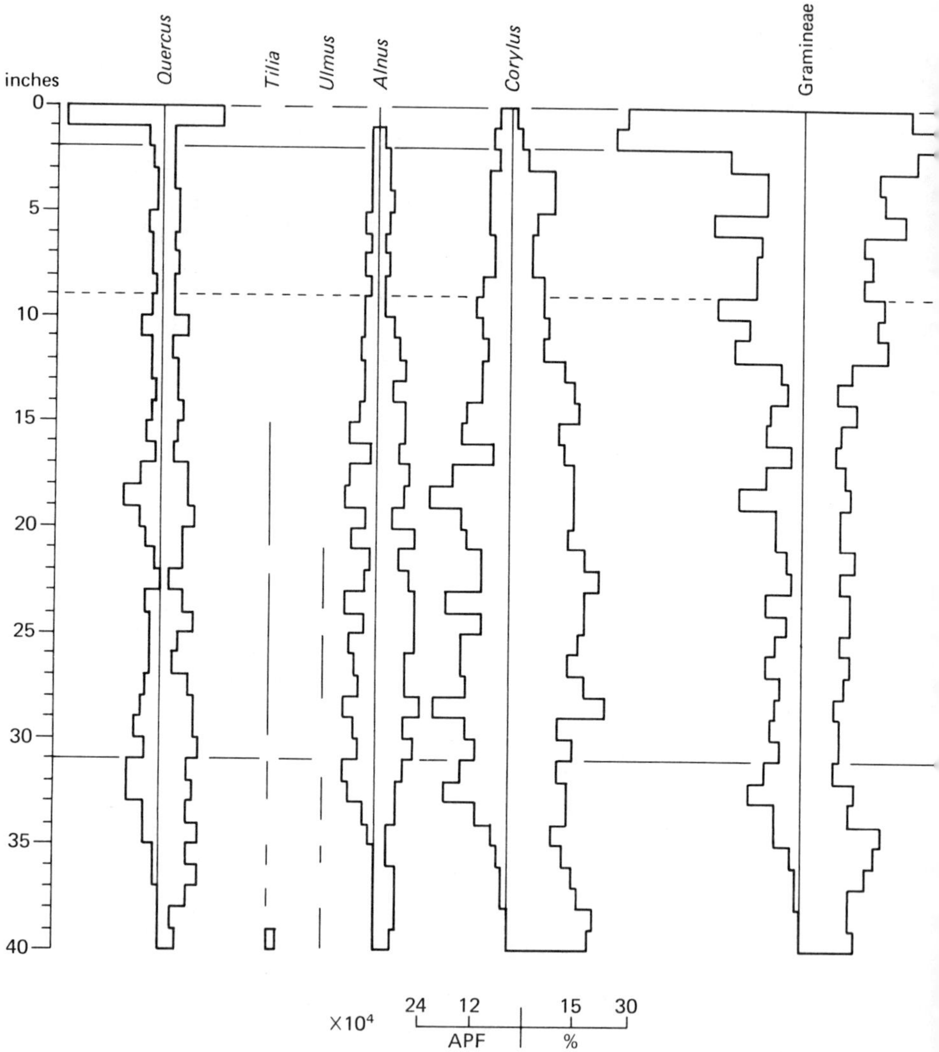

Figure 16. Pollen profile through the mound of a round barrow at Chick's Hill, Dorset. (See also Fig. 23.)

because a supply of earth was needed, or both? Was the construction of an earthwork by the raking up of topsoil a specific technique with a reason? Or was it that the subsoil was too intractable or the work force too weak? And why should material have been brought to a site from a distance? These are questions that can be posed, but not answered by the palynological evidence.

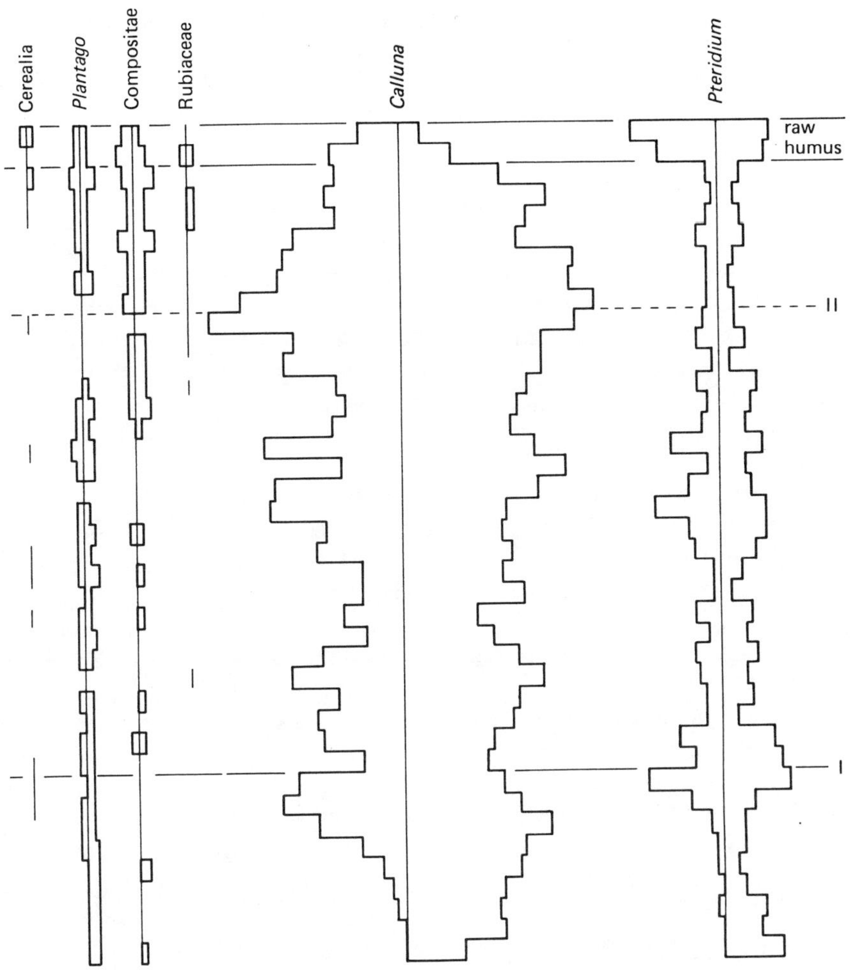

Figure 16. *continued*

Take as an example the barrow at Chick's Hill, Dorset (Ashbee and Dimbleby 1959). The pollen section through this barrow is shown in Figure 16. The 31-inch level is clearly the surface of the buried soil. This shows the characteristic pollen frequency distribution pattern with depth and clearly is an intact profile. The top four samples of this profile are rich in pollen, as shown by the frequency curves on the left side of the histograms for each taxon; below this level the quantities are low. If we examine the pollen frequencies through the mass of the mound up to the 9-inch level it is apparent at once that they are comparable in quantity with the upper layers of the soil, and on the whole the percentages do not vary much from those of the buried soil. The mound was constructed of scraped-up topsoil. At the point of sampling, at any rate, there is no sample which is poor in pollen, so none of this material can have come from below the top 4 inches of the soil profile. We shall return to other features of this profile later.

Another barrow which shows the same pattern is the Fox Howe at Bickley, North Yorkshire (Dimbleby 1962a:Figure 2). In this case some of the frequencies are considerably in excess of the values for the surface of the buried soil. It may be that there has been some truncation of the buried profile, though there are no pedological grounds for thinking so; alternatively the topsoil making up the mound may have come from elsewhere, where pollen was better preserved. The fact that the higher frequencies are also associated with higher percentages especially of tree and shrub pollen, might support this interpretation, or could be due to forest regrowth over the mound after construction.

In contrast with earthworks such as these is the Bronze Age barrow at Portesham (Thompson and Ashbee 1958). It consists of a turf core (which will be referred to again in Chapters 5 and 7), covered with a mantle of unleached gravel. It does not take much imagination to deduce that in an area of podzolized soil the turf of the core came from the soil surface and the mantle from the ditch. Pollen analysis is hardly necessary to prove the point, but it is instructive to see how clear the evidence is. The turf core was sampled for pollen, but the results will not be discussed here; a pollen sequence was also taken from the present surface of the mound down into the earth mantle. In Figure 17 the true surface of the mound is at 4 inches (the reasons for saying this will be returned to below); from this level downwards the frequencies of all taxa, after some fluctuations at the surface, fall away gradually with depth until below 14 inches only traces remains. It is apparent that all the pollen in this section of the diagram has come from the 4-inch surface and that there could have been no pollen present in this material when it was deposited. This confirms what was already obvious, that the material had come from well below the polliniferous topsoil, doubtless from the ditch.

So far we have looked at two simple types of structure—earthworks consisting entirely of scraped-up topsoil, and one of a turf core covered with a mantle of subsoil. Not infrequently, however, earthworks may be made up of layers of material that are visually distinct, either in colour or texture, or both. Sometimes these have been interpreted, quite reasonably, as constructional tip lines, but in other cases they give the impression that the earthwork has been built in stages. In sandy areas colour may be the only key to the layering. Such phenomena had been observed in mounds in the sandy areas of Holland and Germany, and in one series of burial mounds in Germany

Groenman-van Waateringe took samples at key points for pollen analysis. Working on these samples Groenman-van Waateringe was unable to satisfy herself that there were significant differences between the pollen spectra of these mound features and those of the old land surfaces (Groenman-van Waateringe 1979). This does not necessarily prove that there were no stages in construction but shows that there had been no significant change in the vegetation at the time the various materials were receiving pollen. This could be due to a short time interval, perhaps with material from the contemporary land surface being used in construction, or, if there had been a considerable time interval, that the vegetation remained stable over that interval. To investigate the question fully would require more than spot samples; serial sampling would need to be carried out at possible stage junctions.

Visible layering in the sections of earthworks may be produced by various means. One which archaeologists have sometimes failed to distinguish is the horizonation that may develop from pedogenesis from the surface of the earthwork; if this takes the form of podzolization it may produce a dark B_H horizon which can be mistaken for a buried soil surface. But there may also be layering which is real but not apparent to the eye, and this can sometimes be shown up by pollen analysis. In the diagrams which follow, it must be made clear that a horizontal line drawn across the diagram does not represent a horizontal boundary in the field section; it only represents a boundary on the line of sampling. Whether a zone so established extends laterally beyond the sampling line can only be established by further observation in the field. Nor can the pollen tell us anything about the cause of the zonation, though it may be able to tell whether a junction had been exposed to the atmosphere for a length of time.

The first of these two examples is the section of the rampart of Caesar's Camp, Keston, referred to in Chapter 3. There our concern was with the buried soil, in which oak pollen predominated. The pollen profile extended upwards into the base of the rampart (Dimbleby 1962a:77). Here the analyses for these samples differ not only from that of the buried surface, but they vary among themselves. Four layers have been distinguished. The analysis of the lowest one is very similar to that of the buried surface on which it rests, but in the layers above there are noticeable differences. These contain considerable amounts of heather (*Calluna*) pollen, which is absent from the buried surface; the pollen of grasses (Gramineae), another light-demanding taxon, is also more strongly represented. Clearly these materials could not have come from the old land surface immediately adjacent to the point of sampling. Nevertheless, they all agree in having oak (*Quercus*) as the dominant taxon, which suggests that they all come from the same general area.

A very different type of structure has raised similar questions about the

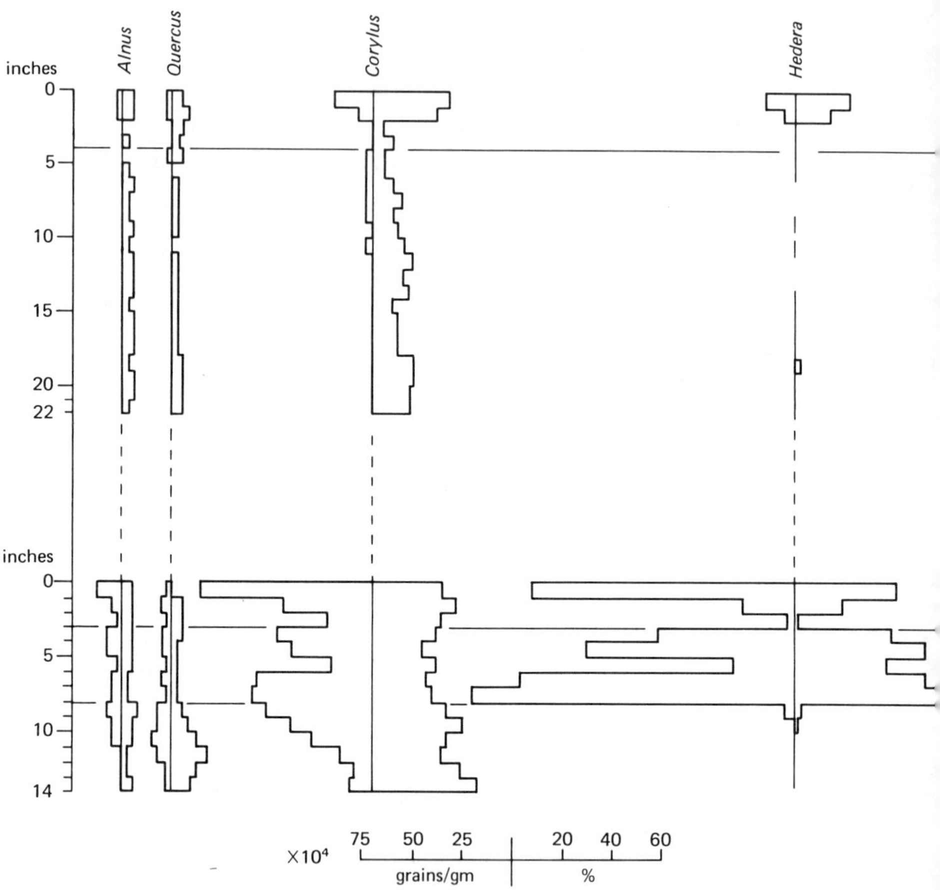

Figure 17. Pollen diagrams of sections through a round barrow at Portesham, Dorset. Upper diagram: section into the mound from present surface. Lower diagram: section through the turf core into old land surface (OLS).

origins of the material of its make-up. A neolithic ossuary at Lutz-en-Dunois (Eure-et-Loir) contained some walling built on an old soil surface. Planchais (1965) carried out pollen analyses on the buried soil and also on the interstitial filling of the walling. Although in this highly calcareous context the pollen was very scarce, she was able to produce spectra which revealed a big difference between the two materials. The pollen from the buried soil was that of a dry open environment, whilst that from the wall filling represented woodland. Both were of Subboreal age. Clearly the wall filling could not have come

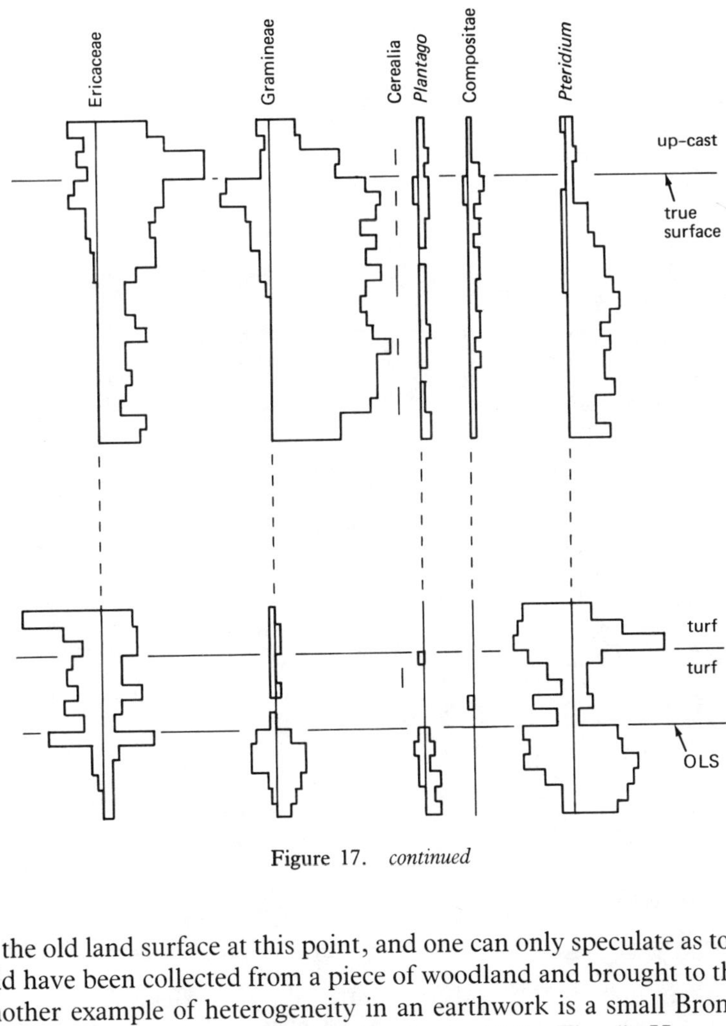

Figure 17. *continued*

from the old land surface at this point, and one can only speculate as to why it should have been collected from a piece of woodland and brought to this site.

Another example of heterogeneity in an earthwork is a small Bronze Age barrow at Burley in the New Forest (Dimbleby 1962a:Fig. 4). Here, too, the buried soil profile is clearly intact, and its surface shows predominantly pollen of *Calluna* and microspores of *Pteridium*, with *Quercus* and *Corylus* pollen also conspicuous. The mound itself can be divided into four pollen layers; the lowest A, 5 inches deep, is an approximate mirror image of the buried soil and as there is no visible evidence of organic structure it is presumably a spit of soil from the ditch. Layer B above it is in complete contrast; at the time I did this analysis I thought I could distinguish *Corylus* and *Myrica* (sweet gale) pollen, but now I am less sure and prefer to label them jointly as 'Coryloid' (Fig. 18). It is 'Coryloid' pollen which dominates this layer, nothing else

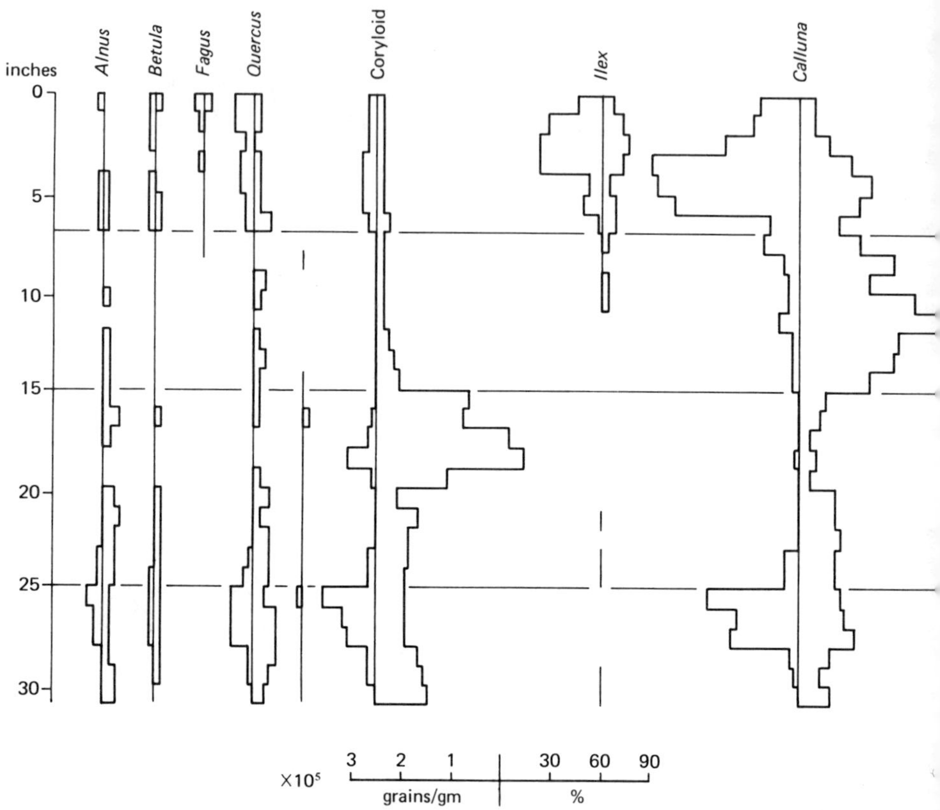

Figure 18. Pollen diagram of a section through a small round barrow at Burley, New Forest, Hampshire.

being present in more than trace quantities. Layer C offers another contrast; 'Coryloid' pollen is virtually absent, and the layer is dominated by *Calluna* pollen. Neither layer A or layer B has any parallel in the buried soil (nor in the overlying material), and it seems clear that the material must have been brought in from a distance. If the 'Coryloid' pollen in layer B is in fact *Myrica*, a bog plant, there is no possible source of it nearer than 200 m. The topmost layer D is rich in pollen, all of which could have been derived from the present barrow surface in the course of its existence since the Late Bronze Age. It is impossible to tell whether there was any pollen inherent in this layer when it was deposited; it would be masked by later pollen. However, the taxa and their distribution with depth—that is, their history—would seem to give the expected vegetative sequences in such an area over such a period of time.

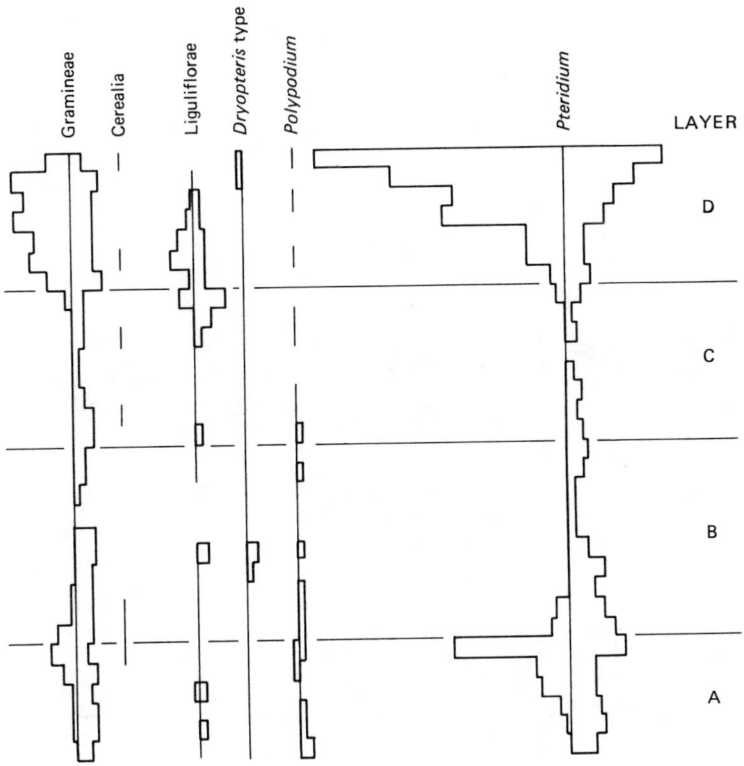

Figure 18. *continued*

It is not only the existence of such layering that is of interest, but also the fact that the boundaries between the layers, as reflected by their pollen content, are so sharply demarcated. The little barrow at Burley stands only 2 feet (60 cm) high at the point of sampling, yet the only substantial evidence of pollen movement is in the active soil on top of the mound, layer D. This observation is not peculiar to this case but seems to be generally true.

An intriguing, but incidental, point in this connection is the pollen curve for the Liguliflorae (dandelion subfamily). At the interface C/D, Liguliflorae pollen shows a noticeable increase; in fact at this interface there is some evidence of downward movement of this pollen type. The suggestion is that at this level a surface has been exposed long enough to allow some Compositae weeds to grow and flower, but strangely the pattern is not exactly matched by any other pollen type. However, we must remind ourselves that we are not looking at layers stretching right across the mound (as far as we know) but only at junctions across a line of sampling.

A parallel zonation, but different in detail, appears in the section of one of

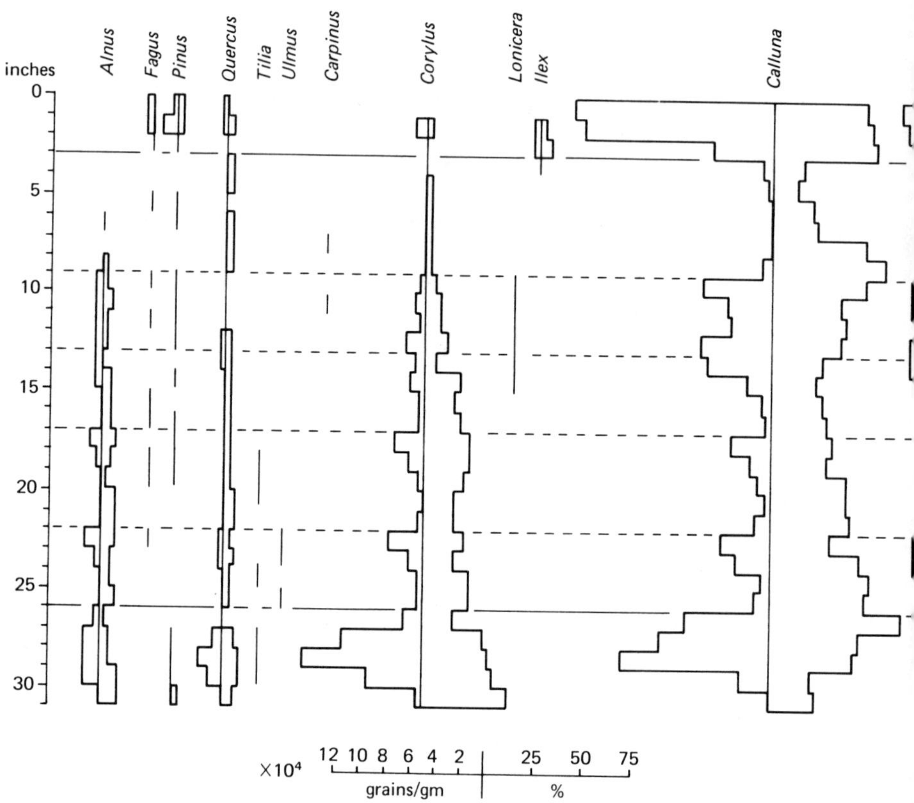

Figure 19. Pollen diagram of a section through a circular bank, Studland Heath, Dorset.

the many circular banks of unknown age (but pre-seventeenth century A.D.) and purpose on Studland Heath (Bowen and Fowler 1963). The interfaces are marked by humus staining and probably represent successive building up of the bank, perhaps as the centre of the circle was periodically cleared out (Fig. 19). Here too there is a pollen type that shows an unexpected distribution: shore weed (*Littorella*), whose characteristic pollen was found at the *base* of each layer. This suggests that it was present in the material being used to build up the bank. Shore weed is a plant that will grow submerged in non-calcareous water, and flowers when the water recedes. The picture is one of standing water in the circular bank; some, but not all of these circular banks, have 'entrances' as this one does, but whether it was a deliberate breach to a previously intact bank one does not know. The archaeological significance of these circular banks is very puzzling; unfortunately, the palynological evidence only makes it more so.

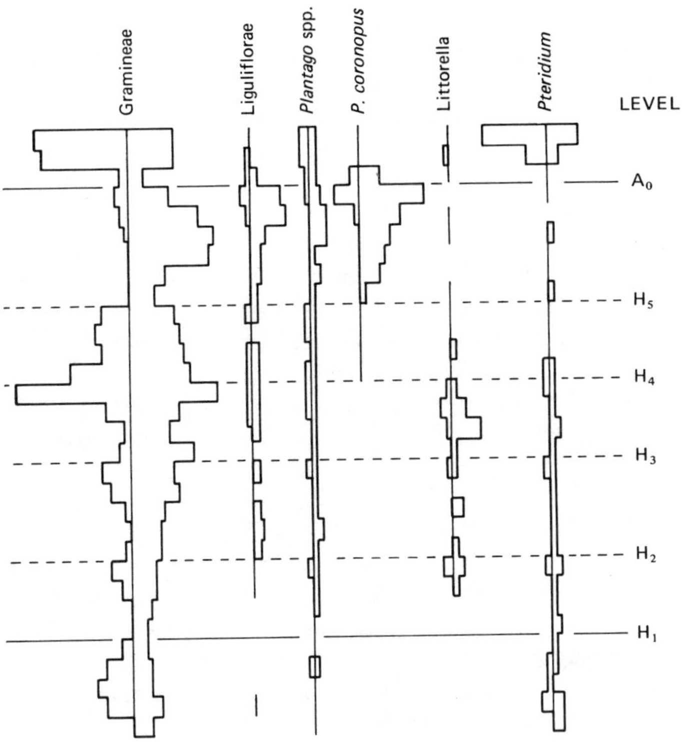

Figure 19. *continued*

PRIMARY VERSUS SECONDARY SURFACES

The question sometimes arises as to whether an earthwork has been increased in size by the addition of extra material, as, for example, a barrow being enlarged after the insertion of a secondary burial. If it were left for a considerable length of time before such an extension was added, one would expect the surface of the earthwork to become vegetated and consequently to develop a pollen profile in the surface layers comparable with that of a soil. One hears suggestions that earthworks whose surface is, say, chalk or gravel may remain unvegetated and conspicuous for a long period of time. The experimental earthworks on Overton Down and at Wareham have shown us that such a condition only persists for a limited period. Within a decade plants are established, even if they do not give total cover, so that a pollen source is created. Apart from this, however, even an unvegetated surface is bound to be receiving pollen from the surrounding vegetation.

In theory, therefore, it should be possible to locate a temporary surface if it has been exposed for a long enough period of time, but I would not like to suggest what this period might be. Sometimes it is quite obvious that there has been such a stage in construction of the earthwork as we see it today. In sections of round barrow on acid soils I have seen clear evidence of podzolization associated with a buried surface in the mound; one such example is illustrated in Plate IV.2 in Dimbleby (1962a). A pollen profile through this section shows that the surface is marked by an increase in the percentage of ericaceous pollen (Dimbleby 1955a). This was an early analysis in which absolute frequencies were not calculated.

Nevertheless, buried surfaces within the body of earthworks can clearly be demonstrated on the basis of pollen alone. Such a case is the so-called Ridge at Lyndhurst in the New Forest, a massive enclosure bank dating from A.D. 1299. Samples were taken through the buried old land surface, clearly visible with its strongly developed heath podzol, and they showed the very high frequency of ericaceous pollen at this surface. The bank itself in the lower part was obviously derived from this soil, but the upper parts were of unleached sand and consequently devoid of pollen. Samples were also taken downwards from the present surface of the bank. The surface itself showed high ericaceous pollen, reflecting the present heathland vegetation, but at 5 inches depth there was a sharp break and a new peak of heath pollen appeared. It is apparent that we have two buried surfaces here, and it may be observed that the lower one is a relatively weak feature and is unlikely to be the primary surface of the bank. There are records of the bank having been repaired in earlier times and it may be that burial of these surfaces is connected with such operations.

In our earlier discussion of the source of materials, reference was made to the topsoil barrow of Chick's Hill (Fig. 16). This section not only shows clearly the position of the old land surface but also what is clearly an earlier surface of the barrow (Level II, at 9 inches), marked by peaks of heather and grass pollen. Above this level the heather becomes progressively less abundant, the grasses and weeds more so, and the tree pollen is diminished. It seems apparent that a further 7-inch depth of mineral soil, again topsoil in origin, has been added to the barrow. In his report the excavator, Paul Ashbee (Ashbee and Dimbleby 1969) gives parallels for composite barrows, built in stages, and remarks that in this case the details would have been undetectable by normal excavation.

One other case of a buried surface is worth mentioning, namely the Bronze Age barrow at Portesham, already referred to as a barrow with a turf core mantled with subsoil gravel. Our interest now concerns the present surface of this barrow. The pollen diagram (Fig. 17) shows unusual features. It is clear that most of the curves show a sharp break at 4 inches, at which level frequency peaks occur for most species followed by progressive falls to negligible values as already described. Clearly the 4-inch level represents a surface. What is the source of the material above it? The pollen features of this shallow overburden are: reduced proportions of grass pollen, much increased values of heather and hazel pollen, and most significantly remarkably high values of ivy (*Hedera*) pollen. These three features are directly paralleled in the buried soil and the sods of the turf core; ivy in particular, which usually occurs as mere traces, being an insect-pollinated species, was present at the old land surface in extraordinary abundance. At the time the report was published on this barrow no explanation could be found for this, but now possible explanations can be offered and will be presented in some detail in Chapter 7. In the present context pollen not only served to demonstrate that a surface had been buried but through what might be termed a 'marker' species in it the source of the overburden could be established. It seems quite clear that the top 5 inches of earth had been contaminated with, if not directly derived from, material excavated by a previous sondage which reached the core of the barrow.

In later chapters, particularly Chapter 7, it will be argued that some high pollen concentrations may in fact be artifacts. Where pollen of entomophiloous species is grossly overrepresented it seems impossible that it is in equilibrium with the pollen rain. In the case of *Hedera,* as we shall see, it seems likely that it was being used as a fodder plant which may have been stored on the site, or the concentrations we see may have been the result of accumulation in animal dung.

Whatever the explanation, it may be important to realize that the pollen may have been superimposed on the usual distribution pattern that we are now familiar with. An example where this distinction is important is the old land surface under the same round barrow at Portesham (Fig. 17). On the assumption that the actual level of the old land surface would be marked by a peak in the curves of at least one of the taxa, this level was originally taken to be at sample 8 (Thompson and Ashbee 1958), where *Hedera* reached a clear peak. However, if it is accepted that the *Hedera* pollen may be derived from material added to the site, then the *Hedera* curve must be discounted. When this is done the true surface is seen as at sample 9; moreover, the pollen curves show that the Ericaceae had recently increased in importance, though still outnumbered by the pollen of woodland plants, mainly hazel. The picture one gets is of a developing clearing in a woodland background. Not only does this make better sense ecologically, but it accounts for the discrepancy, inexplicable by the previous interpretation, that a number of the other curves show a sharp break between sample 9, now thought to be the true surface, and sample 8, which now seems to be part of a turf or other addition to the surface.

It was mentioned above that buried surfaces can sometimes be recognized by the soil horizonation that had developed in association with them. This question figured strongly in the discussion on the Chick's Hill barrow, and indeed such features, where recognized during excavation, could be of considerable interest to the pedologist, particularly in relation to the chronology of profile development. The distinction between topsoil and subsoil barrows seems to be important in determining the speed of soil development; development is slower in subsoil materials, whereas topsoil seems to be predisposed to rapid iron movement and resultant horizonation (Dimbleby 1962a:16).

DITCHES

It is often assumed that ditches are the most promising feature of an archaeological site for pollen analysis. This seems to be based on an assumption that conditions in them approximate to those in a wet peat deposit where good pollen preservation may be expected. If a ditch is permanently waterlogged, pollen may be preserved, though the interpretation of that pollen may present problems. If a ditch is only damp, or is only seasonally wet, the deposits will be aerated, at least during the seasons of highest temperature, so that microbiological attack may be expected.

More serious, however, is the question of the origin of any pollen found in a ditch. The experimental earthworks on Overton Down and Wareham have given us valuable information on the complex processes involved in the 'silting-up' of a ditch. Taking the case of a ditch cut into a soil acid enough to preserve pollen, we can expect that the lowest deposit in the ditch will contain a mixture of pollen from the contemporary topsoil, together with pollen which may wash or blow in from the lower layers of the soil; as we saw in Chapter 1 wide differences in pollen assemblages can occur over a few inches' depth in a soil profile. Added to this is the continuous pollen rain which will fall into the ditch year by year, from the plants growing in the ditch as well as from the general vegetation of the area. There is thus an admixture of contemporary pollen and pollen of greater age; usually the same taxa are involved (grasses, weeds, hazel, bracken, etc.) but in varying proportion.

Whether there is any value in carying out pollen analysis on a ditch depends to some extent upon what kind of information it is hoped to obtain. If it is purely for dating there are, as we have seen, sources of error which must render the technique open to question. On the other hand, depending on the nature of the ditch fill, some information may be obtainable about the changes in the local vegetation as the ditch filled up. Even this, however, is not always clear-cut. Figure 20 shows the diagram published by Tinsley and Smith (1974) of a pollen sequence through a Romano-British ditch at Fortress Dike, Yorkshire. It is one of the few ditch pollen profiles published. At the

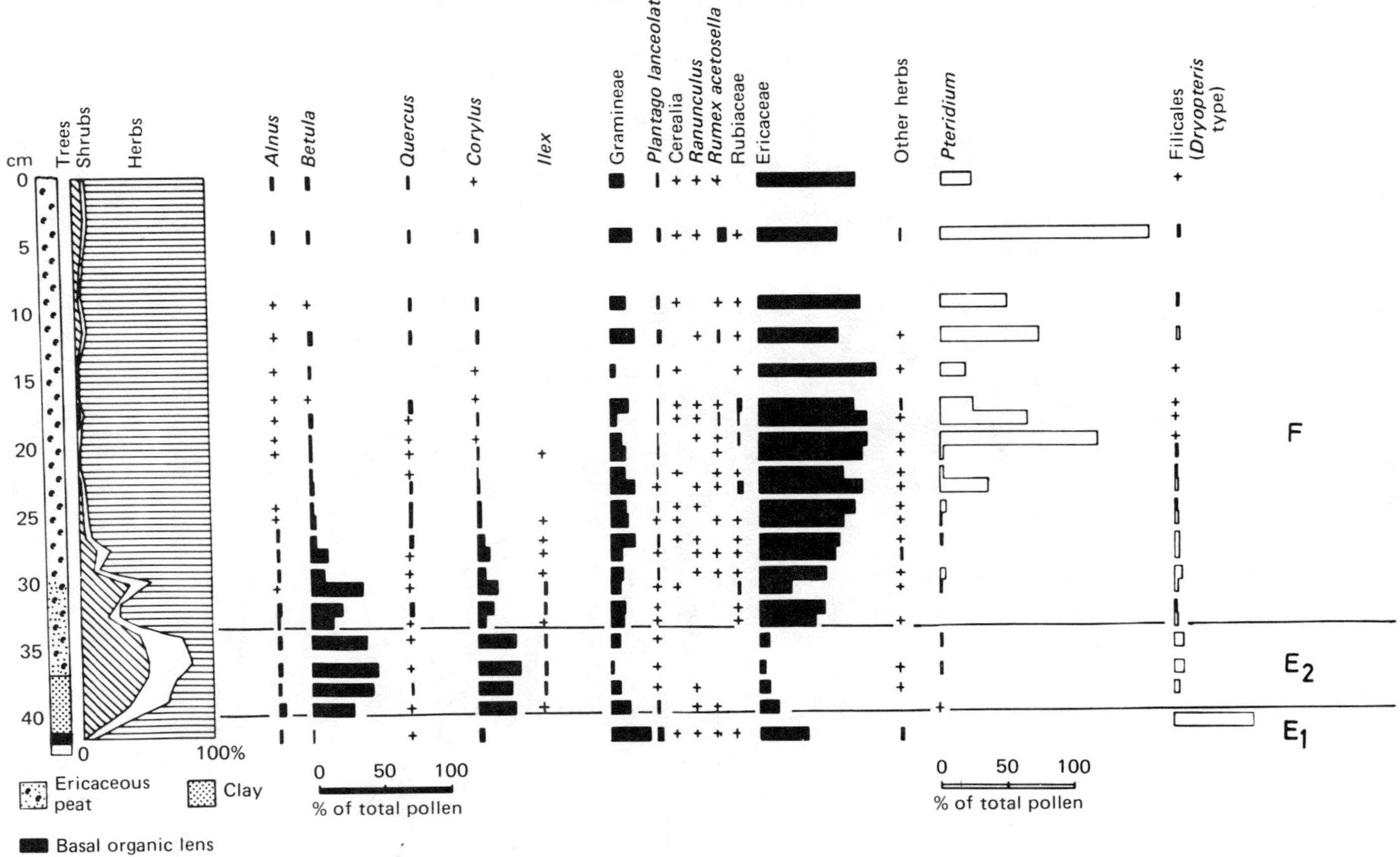

Figure 20. Fortress Dike, Yorkshire. Pollen diagram of ditch fill. (From Tinsley and Smith 1974.)

very base of the profile there is an organic lens which gave a pollen spectrum virtually devoid of trees and dominated by grasses and heath species. Whether this lens formed *in situ* or was derived from organic material from the lip of the ditch, it comes closest of all the samples to representing the landscape at the time the ditch was cut. Above this is a layer of clay 5 cm thick containing abundant pollen of birch and hazel, and this in turn gives way to ericaceous peat in which the tree pollen progressively falls away, to be replaced by the pollen of Ericaceae plus high levels of bracken spores.

It seems incontrovertible that the upper part of the profile represents the development of heath or moorland with bracken becoming a major component in a way that it was not at the time the ditch was cut. A question must hang, however, over the tree pollen in the lower part of the ditch fill; tree pollen was not significantly present in the basal organic lens itself. The authors interpret the clay layer above the organic lens as 'wash from upcast.' Fortunately they also publish a pollen diagram of a section through the bank, and it is apparent from this that the tree pollen in the ditch fill could indeed have been derived, in part at least, from erosion of the bank itself, as the authors acknowledge. It does not follow that because the rest of the ditch fill is ericaceous peat that it is not contaminated by pollen from the mineral soil. On the North York Moors I have found living moss polsters to contain pollen of hazel and lime, species long absent from the area. This has clearly been derived from tracks which have broken through the soil surface and exposed the A_2 horizon which contains these old pollen types. Whether Tinsley and Smith are right or not, and not having been involved in this work I hesitate to give an opinion, there are obviously a number of factors which have to be taken into account in interpreting pollen profiles from such ditch fills.

The larger a ditch the less it will be influenced by contamination from its sides, but the question remains whether pollen could be brought in with the material filling the ditch. The waterlogged ditch of the Roman fort at Bearsden, Scotland, was investigated botanically by Dickson *et al.* (1979) and also biochemically (Knights *et al.* 1983), showing a distinct layering in the ditch deposits. The deepest layers contained a cultural pollen suite, together with fragments of food plants, whilst the plant remains in the upper layers suggested that the ditch had become overgrown by shallow-water plants and ultimately shrubs. Biochemical analysis of the deeper deposits showed that they contained sewage. Human and animal faeces can contain pollen (see Ch. 7), so it cannot be assumed that all the pollen in such a ditch is derived from the surrounding countryside or any vegetation in the ditch itself.

In this case the abandonment of the fort led to the ditch carrying its own vegetation. In a similar way non-waterlogged ditches may also become stabilized and overgrown, and though I am not aware of any pollen analysis on such a stabilized ditch surface it could give results which are interpretable. In many silted-up ditches there is a clear horizon where a soil has developed, indication a long period of stability at that level, and with a series of samples through such a profile it may be possible to distinguish the recent pollen from background pollen in the matrix by their distribution curves. Difficulties arise, of course, where the same species are present in both categories, but even here the state of preservation can sometimes help to distinguish them. However, whilst this may provide useful information about the vegetation history of the area it would hardly interest the excavator of the original earthwork; it could postdate his monument by many centuries.

LYNCHETS

Whether lynchets should be regarded as archaeological structures must depend on one's view on how they were formed. Strip lynchets would probably qualify in the eyes of most archaeologists, but the lynchets demarcating the so-called Celtic fields are regarded by some as the unavoidable result of slope cultivation, whilst others see an element of deliberate construction in them. However, lynchets are prominent archaeological earthworks, whether accidental or not, and as such might be the object of palynological study. Insofar as they have been investigated biologically, this has been mainly by mollusc analysis; as the most dramatic examples are on chalk soils this is the obvious tool. I myself on one occasion tried to locate an old land surface beneath a lynchet on Lullington Heath, Sussex, using pollen analysis. Pollen preservation was poor and the exercise failed for the reason given earlier (Ch. 3), that ploughwash overlying a plough soil obscures any recognition of where a surface might lie.

As far as I am aware, no attempt has been made to examine lynchets on noncalcareous soils using pollen analysis. It would seem to be worth trying, particularly for a large lynchet where there might be the possibility of the preservation of an old land surface. The body of the lynchet, too, could reward investigation, both by pollen analysis and by seed analysis. It would be valuable to find botanical evidence of weeds—and therefore perhaps of farming practices—to see if they could be correlated with lynchet formation. At the Butser Iron Age Farm experiment observations are being made which could be linked to such an enquiry (Reynolds 1976) though once again, of course, this experiment is on chalk and therefore less favourable for palynology.

POSTHOLES

As with ditch fills, archaeologists sometimes see postholes as sources of pollen from which dating or environmental information may be obtained. The reasons for this belief are presumably the same as for ditches—that the fills are rich in organic matter, and therefore should preserve pollen. This impression is probably most strongly obtained with calcareous soils, where the fill, derived from the A horizon of rendzinas, contrasts strongly with the staring white of the parent material into which the posthole has been cut. In fact, as we have already seen, the organic layers of rendzinas are highly active microbiologically, and pollen is only poorly preserved. The fact that the mull layer of a rendzina and a peat both have a dark humic colour is no indication that the two have similar properties of preservation. Actually their organic content derives from quite different circumstances in the two cases, and their physical and biological properties are completely different. In any case, much of the organic material in a posthole may be derived from the rotting of the post itself and will therefore be irrelevant palynologically.

On the face of it there is no reason why the fill of a posthole should give a contemporary pollen spectrum, and there are a number of reasons why it might not. Once the hole has been dug and the post inserted, the material used for packing—which eventually becomes the fill, added to by collapse from the sides—may or may not be derived from the topsoil of the surrounding land surface. It is perhaps more likely to come from the mineral layers of the soil which, having a greater clay content, bind better; such material is likely to have a different pollen suite from that of the topsoil in equilibrium with the contemporary pollen rain.

I have come across very few examples in the literature of pollen analyses from posthole fills. Groenman-van Waateringe (1979) took samples from what she described as a post trench, as part of a suite of samples from various contexts in a German round barrow. Heim (1979) also included a sample from a stakehole amongst a series of samples from a variety of situations. Neither of these authors, however, questioned the validity of holes or trenches as a source of reliable pollen spectra, but simple used them because they contained pollen.

For my part, I have always avoided as far as possible sampling from postholes because of the doubts already expressed. On one occasion, however, I did carry out some analyses on samples supplied by an archaeologist, of which two samples turned out to be from postholes. The site was the Iron Age hill fort of Ranscombe, Sussex (Table 2). Six samples were submitted, three from the Iron Age old land surface, two from contemporary postholes, and one from the modern land surface. (This last will be discussed again in Chapter 7.)

Table 2 shows a condensed form of the analyses; it will be seen that in both the old land surface and in the postholes the pollen frequency was low. There are some strong parallels between the two analyses from the Iron Age contexts. In both, the pollen of woody species is scarce; the slightly higher value for tree pollen in the turf is largely accounted for by one anomalously high count of *Betula* (birch) in one sample. Many pollen analysts omit *Betula* from

Table 2
POLLEN CONTENT OF SAMPLES FROM AN IRON AGE HILL FORT AT RANSCOMBE, SUSSEX

Taxon	Iron Age Turf	Iron Age Postholes	Modern Turf
	% OF TOTAL POLLEN PLUS FERN SPORES		
Tree pollen	4.8	3.2	14.0
Corylus	1.5	3.9	—
Gramineae	3.4	5.5	15.3
Cerealia	0.1	—	20.3
Liguliflorae	20.1	18.9	17.1
Pteridium	63.6	55.4	7.2
	GRAINS/GM SOIL		
Excluding *Betula*	1692	1077	590
Absolute pollen frequency	8405	8460	27,750
NAP/AP %	552	993	558

the tree pollen totals because of its high pollen production and its erratic and spontaneous occurrence in the landscape. Otherwise the two analyses are dominated by Liguliflorae and bracken (*Pteridium*). It could be argued that the two spectra are sufficiently similar to indicate identity. I hesitate to accept this interpretation for several reasons.

There is a marked difference in the NAP/AP ratio, suggesting that the posthole material came from a less wooded environment; but if *Betula* is excluded, the turf is seen to have a much higher ratio than the posthole, indicating progressive clearance. The old land surface also contains less hazel (*Corylus*) pollen; the progressive disappearance of hazel is a feature of the final stages of the replacement of woodland by an agricultural landscape, so some lapse of time may be indicated.

There is also a difference in the percentages of bracken (*Pteridium*), one of the predominant taxa. This may not be significant, but the difference is substantial, and if bracken spores are to be seen as arising from agricultural manuring (Ch. 7), it might reflect the technological development of farming through the Iron Age, as described by Fowler (1983).

It has to be recognized that the man-made landscape of the prehistoric period has few ecological indicators by which it can be dated, even relatively. It would therefore be rash to assume that these spectra are identical when there are some important differences between them, differences which could represent the passage of centuries. On this limited evidence I remain unconvinced that postholes can provide pollen spectra that can be accepted unequivocally as contemporary with the cutting of the postholes.

5
OPEN SITES

INTRODUCTION

It is very easy to imagine the surface of the soil as a fixed datum which has been changeless over the millennia. A metaphor I have used elsewhere is that the landscape is a patterned carpet over which man has tramped for thousands of years. In broad terms this may be true, but when we begin to look at the land surface itself it is apparent that we need to treat this image with care and some reservations. It always surprises the uninitiated, seeing an excavation in the centre of one of our older cities, that the Roman or Saxon levels are many feet below the modern road surface; it needs an appreciation of the way man lives and the building materials that he brings into a settlement site, to understand how over the centuries such aggradation can occur.

When it comes to the rural areas the operation of such factors is not apparent. Palynology of soils, however, has shown that even here all is far from changeless. A long time age I published a paper (Dimbleby 1961b) in which it was shown that out of about 40 soil profiles that had undergone pollen analysis at that date, about a third of them showed evidence of a buried soil within the profile. These were not archaeological sites (one or two of them were *beneath* archaeological sites), so that the burial of an old surface was not to be seen as a deliberate activity of man—though it may often have been an inadvertent result of his use of land. If a third of these soils showed aggradation, it may be inferred that there should be a third which had lost soil material—much less easy to demonstrate—leaving only a third unaltered in this respect. Over the years I have had no cause to alter the general impression given by that early paper, though I would not guarantee that the proportions ratios are exactly as suggested there.

EXAMPLES OF BURIED SOILS

The burial of archaeological sites by various agencies has already been discussed, and the point to be emphasized here is that these processes do not

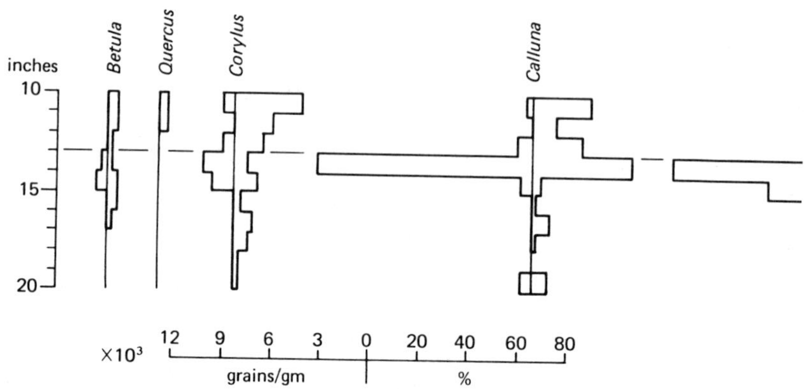

Figure 21. Pollen sequence through a buried level at Cock Heads, North York Moors.

only affect man's sites but are much more widespread and are not invariably related to human activity. In fact a number of these examples occurred long before the Neolithic, when cultivation became established; a number of mesolithic sites will be discussed below, but even here we cannot rule out some element of anthropological influence.

One of the most striking examples of a buried soil, at Cock Heads, has already been referred to. Here a podzol profile contains within it an earlier soil whose surface is marked by high frequencies of Gramineae, *Calluna*, and to a lesser extent, *Corylus* (Dimbleby 1961b). Distributed through this soil are traces of Late-glacial taxa (Fig. 21). This soil was overwhelmed, probably by hillwash, with a deposit in which later pollen history can be read; in due course this layer became peat-covered (Fig. 22).

Two other examples occurred beneath round barrows. At Chick's Hill, already described in detail, a remarkable pollen assemblage was found beneath the old land surface whose pollen analysis is shown in Figure 23. This earlier buried level was marked particularly by the sudden abundance of lime (*Tilia*), both in frequency and in percentage. The pollen data for this level are given in Dimbleby (1962a:Figure 10) and (Dimbleby 1961b:17 and Figure 1). The remarkable feature of this site is that it was on the top of a small hill, which poses a problem regarding the origin of the gravel which covered this buried level.

The other example mentioned was at Reasty Top on the Tabular Hills in North Yorkshire, where a buried level, similarly dominated by lime but with few associated taxa, was found beneath the old land surface of a small round barrow (Dimbleby 1962a:101). It is not possible to put a date to either of these buried levels; though they are clearly much earlier than the old land

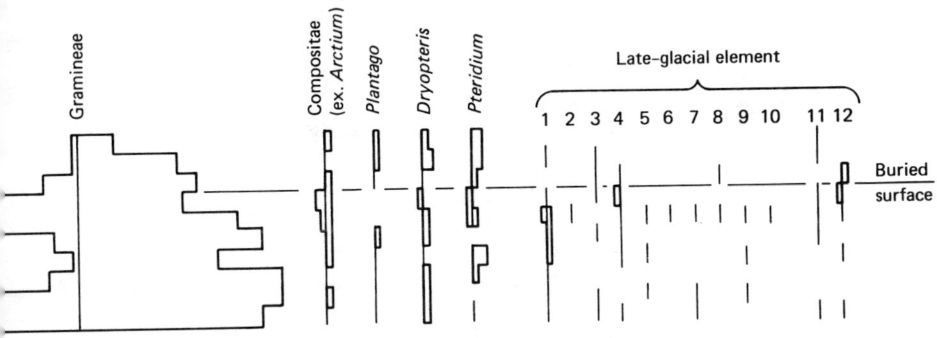

1. cf. *Arctium*
2. *Armeria*
3. Caryophyllaceae
4. Onagraceae
5. *Polemonium*
6. *Polygonum bistorta*
7. Rubiaceae
8. *Succisa*
9. *Thalictrum*
10. *Valeriana*
11. *Botrychium*
12. *Lycopodium selago*

Figure 21. *continued*

surfaces of the barrows, which in each case have had time to form a well-developed pollen profile, it could not be asserted in either case that the buried level occurred in primary forest. In other words human influence cannot be ruled out.

To go on to sites not known to be associated with archaeological remains of any sort, two other examples may be quoted to illustrate the instability of soil surfaces, these both clearly being post-neolithic. The first of these is a heathland soil at Burley in the New Forest. Here the pollen profile is in three clear sections (Fig. 24). At the base is a forest-dominated pollen assemblage, with lime (*Tilia*) featuring prominently. Above this is a zone of six 1-inch samples, whose pollen spectra fluctuate wildly. This zone clearly represents much disturbed material, probably moved and redeposited by soil creep. It does not have the appearance of disturbance by cultivation, which results in more complete mixing of the inherent pollen over a greater depth than these separate 1-inch-depth samples. Above this heterogeneous layer is an 8-inch depth of soil which displays a typical undisturbed pollen profile, showing changes ranging from cultivation among hazel scrub at the base to heather-dominated heath at the surface. This example has been quoted to indicate that some episode of disturbance can be recognized between the lower and the upper zones.

More usually two zones of contrasting pollen content are separated merely by a sharp boundary. An example of this is the soil of Pondhead Inclosure, in the New Forest (Fig. 5), has already been described. The diagram has been redrawn, with *Corylus* and *Myrica* amalgamated as "Coryloid." This does not

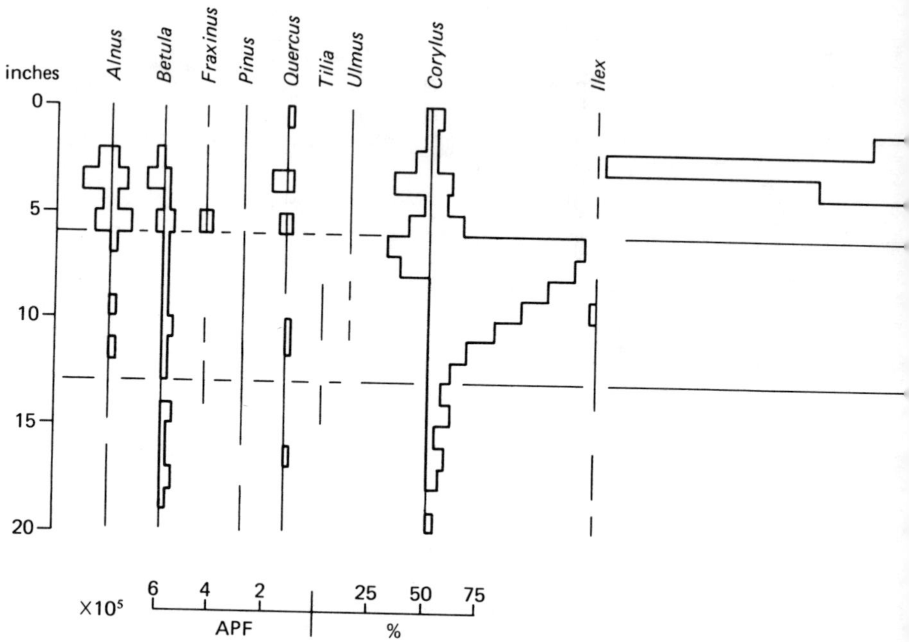

Figure 22. Pollen diagram of a complete soil profile from Cock Heads, North York Moors.

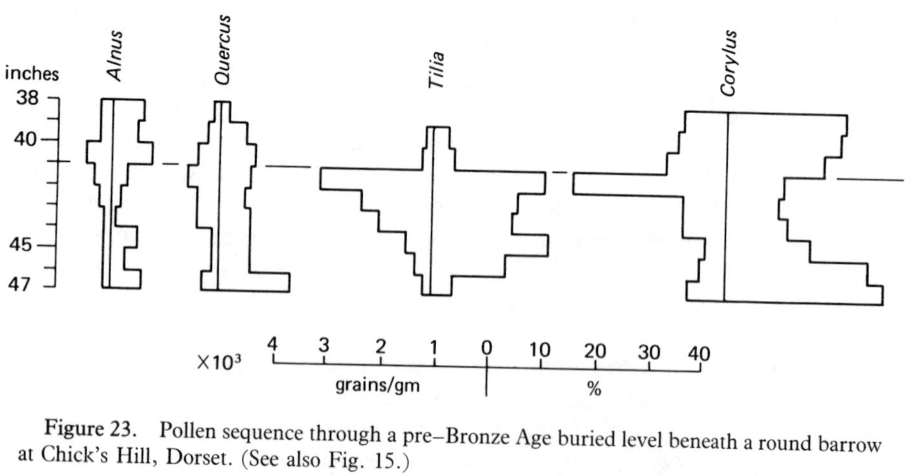

Figure 23. Pollen sequence through a pre–Bronze Age buried level beneath a round barrow at Chick's Hill, Dorset. (See also Fig. 15.)

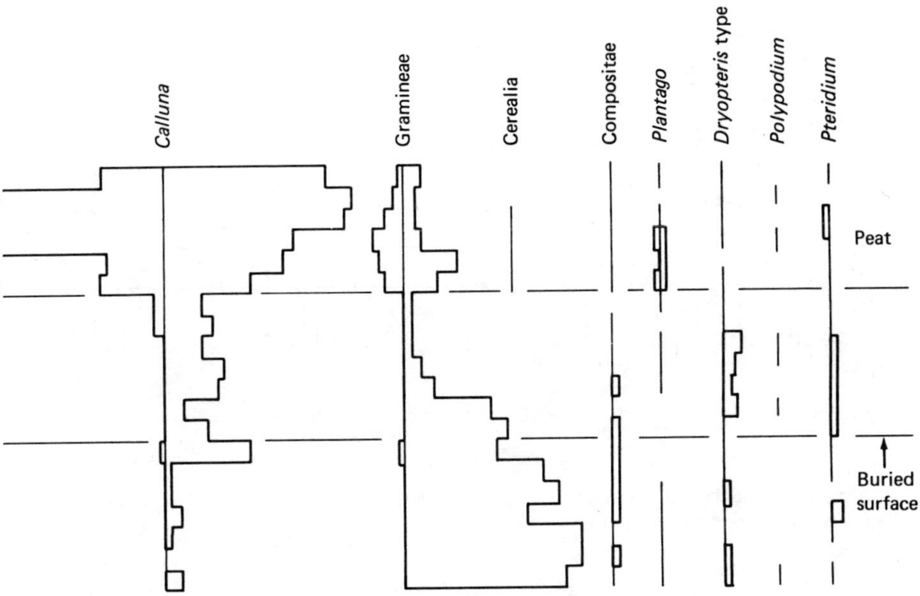

Figure 22. *continued*

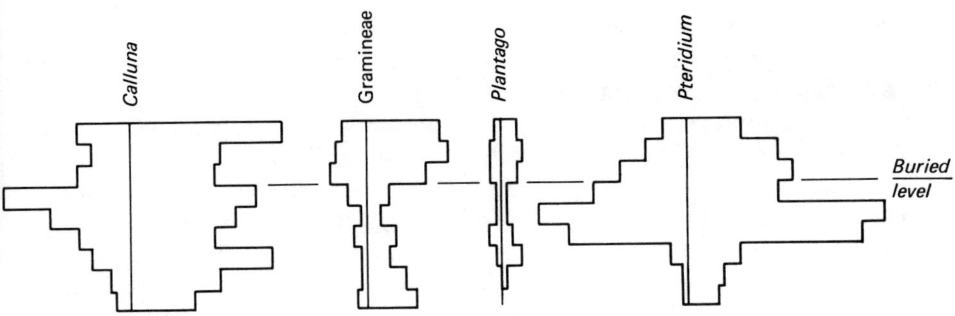

Figure 23. *continued*

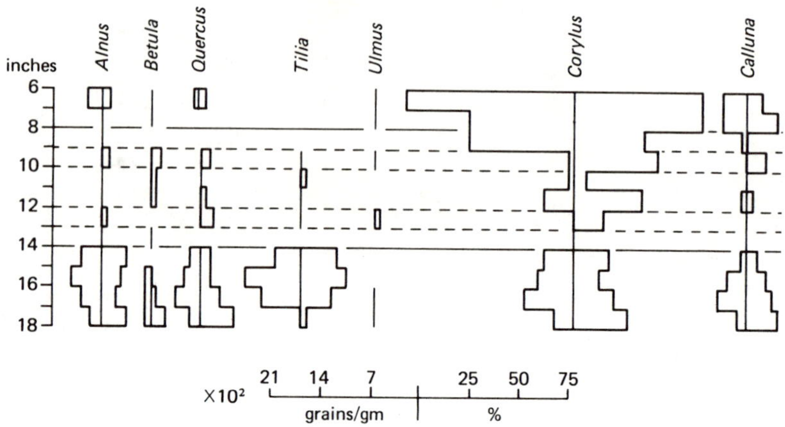

Figure 24. Part of the pollen profile through a heath soil at Burley, New Forest, Hampshire, showing disruption by soil movement. C, Cerealia.

affect the recognition of a buried surface, but of course it makes ecological interpretation more speculative.

The soil pollen profile falls into two zones. There is a lower zone of closed forest, with oak, lime, and alder predominant, contrasting with an upper zone in which a more open type of forest is converted, *via* an agricultural phase, to heathland (and subsequently planted up as woodland). A more detailed analysis is given in Dimbleby (1962a:95). In this profile there is a markedly greater preservation of pollen in the upper zone than in the lower one, a common but not invariable feature of such profiles, which will be seen again in some of the archaeological sections examined below. No doubt this reflects the progressive acidification of these noncalcareous soils, which will have the effect of reducing pollen decay.

These examples serve to show that the phenomenon of buried levels in soil profiles, as revealed by pollen analysis, is not uncommon and can occur unconnected with archaeological sites. Moreover, buried levels occur in all parts of country that have been investigated in this way and they occur in a variety of soil materials. The majority of examples known are in sands, but this may be because sandy soils are very favourable for pollen analysis and have received a large share of attention. In Finland Vuorela (1982) has also obtained pollen evidence of a buried level, apparently covered over by a 30–40 cm depth of blown sand, which she attributed to prehistoric cultivation.

In this chapter the term *open sites* means sites which are not covered by earthworks of any sort. It is a moot point whether sites beneath a peat deposit, or even a mobile dune, should not be regarded as open sites but the decision

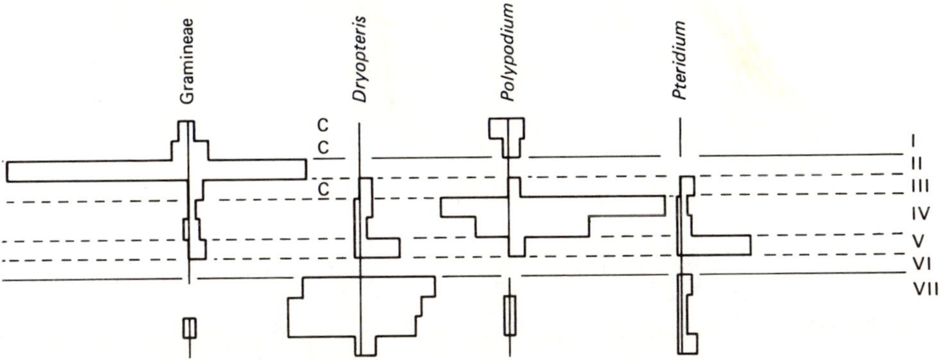

I Normal pollen sequences from present surface.
II Dominated by Gramineae and *Corylus*.
III Dominated by *Corylus* and trees.
IV Dominated by *Polypodium*.
V Dominated by *Pteridium* and *Dryopteris*.
VI Pollen almost absent.
VII Dominated by trees, *Corylus*, and *Dryopteris*.

Figure 24. *continued*

was arbitrarily taken to treat them as buried land surfaces. By doing so it allows this chapter to concentrate on the remarkable group of microlithic sites widely found on heathland and moorland, which have a certain degree of consistency in their features.

Before turning to these, however, the neolithic site at Rackham, Sussex (Holden and Bradley 1975; Dimbleby and Bradley 1975) must be considered as a site in which burial of artifacts was by earthworms, not by mass movement of soil. The pollen diagram is shown in Figure 25. From the column at the right-hand side of the diagram it will be seen that the artifacts occurred at a depth of 15–20 cm in what is now a strongly leached podzol. They consisted of a rich assemblage of neolithic flints, associated with charcoal, and the site was probably associated with butchery and the preparation of animal materials.

The pollen diagram shows no good evidence of a sharp break, covering both frequencies and percentages, in the curves for a number of taxa; contrast with Pondhead, above (Fig. 5). However, the diagram does show distinct trends. From 17.5 to 30 cm it is largely dominated by tree pollen—oak and hazel with subsidiary amounts of alder, lime, and holly (*Ilex*). Moving up the diagram the tree pollen diminished and grass and heather pollen take over until *Calluna* (heather) reaches a strong peak. As this falls away, grasses increase, but the spectrum remains one of open heath conditions. Interpreting this ecological sequence now in terms of soil, one might expect that the deciduous forest phase would be associated with a brown earth type of soil in which earthworms were present. As the forest was removed, perhaps from

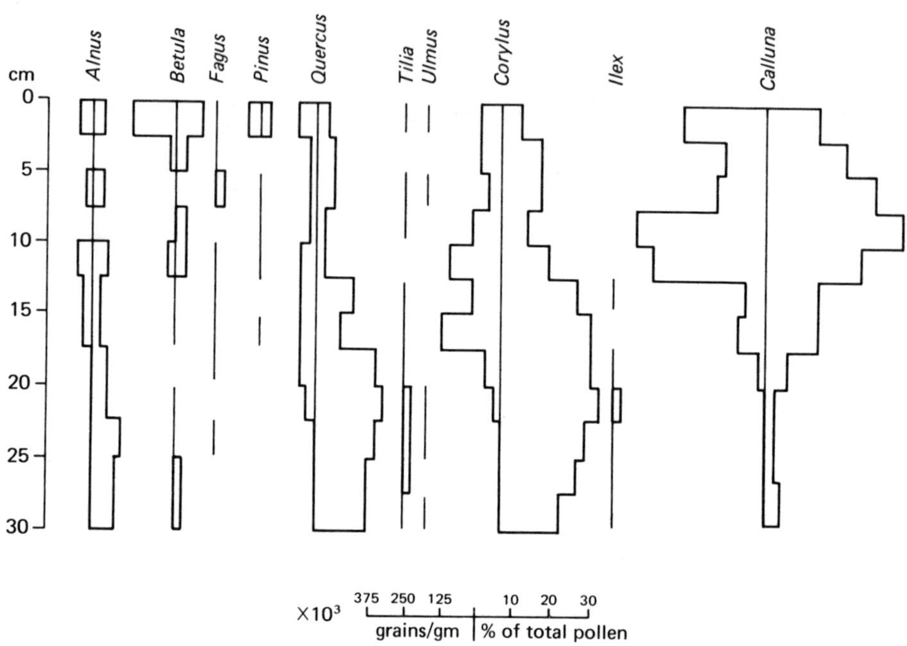

Figure 25. Pollen profile through a neolithic site at Rackham, Sussex.

the Bronze Age onwards, the resultant death of the tree roots would mean that bases would no longer be brought up from the subsoil, with the result that the A horizon would become more acidic. This trend would no doubt be aggravated by the regular use of fire. Acidification would progressively eliminate the earthworm population and the soil would develop into the heath podzol seen today.

Where do the artifacts fit into all this? Since all the pollen in this soil has apparently originated from what is now the present surface, it must be assumed that the artifacts were also at that level of originally. Their movement to their present position would seem to be a classic example of the movement of a whole occupation floor *en bloc* under the influence of earthworm action in much the same way that Darwin (1881) observed the burial of a whole tesselated floor. All objects of a diameter of over 2 mm, which are too big for the earthworms to ingest, will be progressively buried by the fine material which is ingested and cast at a higher level. It is apparent that in the course of time, burial to the observed level could have taken place. But what of the pollen? It would have been ingested with the fine earth, with the result that in the

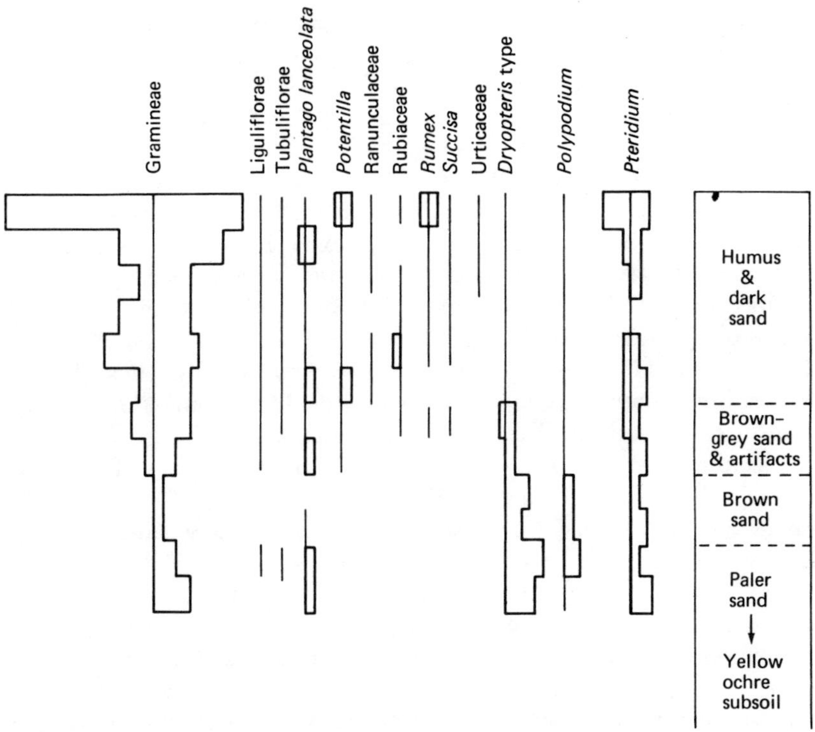

Figure 25. *continued*

course of time it would be uniformly distributed through the layer of worm action. Once the soil had become too acid for earthworms, however, there would be a change in the process of pollen distribution as already described in Chapter 1. The deciduous forest pollen would progressively move deeper, and the pollen of the new ecological dominants, the grasses and heather, would start their slow progression downwards, no longer subject to the disturbance from earthworms.

The reason for taking this example in some detail is to show that in a heathland soil the agency of earthworm burial has once been present, but that the end result of its operation is to bury the artifacts without any clean break developing in the pollen curves. This has to be established before we can go on to consider the microlithic sites referred to above.

OPEN MICROLITHIC SITES

As the mesolithic people did not construct earthworks, their sites come within our category of open sites. Many of them occur on heathland, not necessarily because acid soils were preferred (and as will be suggested below, the soils were not necessarily as acid in mesolithic times). Acid soils have been of less value for agriculture than more fertile soils so that open sites have persisted, being less subject to disturbance by the plough.

However, the pollen analyses of microlithic sites on acid, usually sandy, soils often show a complexity that would be undetected by visual examination of exposed sections in excavations. As with the unconformities in the soils described above, breaks often come in the A horizon of the podzol profile, where any soil structure has disappeared, or in the B horizon where the deposition of iron and humus compounds may obliterate any possible differences of colour or texture.

Under this heading I want to discuss six southern English sites which have these features, though no two of them are exactly the same. They are Iping (West Sussex), Minsted (West Sussex), Oakhanger VII (Hampshire), Longmoor (Hampshire), Winfrith (Dorset), and Addington (Kent). All of these lie on acid sandy soils, four of them in the Weald. A number of similar sites have been investigated archaeologically (e.g., Shennan and Schadla Hall 1981), but only these six have been subjected to pollen analysis. All six were excavated by different archaeologists, each one asking me to carry out an *ad hoc* pollen analysis. This may help to explain the considerable variation in the presentation of the pollen data. They were not studied as a series, and no previous attempt has been made to make comparative studies of the six sites. Literature references will be given as each site is described and discussed, and from these it will be seen that some were excavated over 20 years ago. Some of the published reports describe microliths as occurring in discrete layers in the soil; in fact in one case the flints are described as being concentrated at a definite level—a 'floor'. More recent reappraisal by Jacobi (personal communication, 1982) casts some doubt on the apparent stratification of the flints; Jacobi has himself excavated a new site (Longmoor), paying particular attention to the vertical distribution of the artifacts in the soil profile, and this is one of the sites to be discussed here from the palynological point of view. On the basis of his studies Jacobi feels bound to question the reported occurrence of the microliths in discrete layers. In one or two cases artifacts do not occur in the topmost layer of the soil, but apart from these they are distributed without a break throughout the depth of the site. This is not to say that there is not a variation in abundance at different layers in the profile; certainly the artifacts in some cases occur in concentrations at certain levels and one can see why they should have been interpreted as forming 'floors'. The histograms published by Palmer and Dimbleby (1979:Figure 10) for Winfrith illustrate this point graphically. I will return later to the question of how flint artifacts could have become distributed through a depth of soil, but at this point I am prepared to accept Jacobi's view that in these sites artifacts are not normally concentrated in 'floors' or occupation levels. In view of this, and the unevenness of archaeological recording on some of these sites, I propose to ignore the flint distribution in interpreting the pollen diagrams. As we shall see, it is not necessarily a case of omitting relevant information, for there is good evidence that in most cases the pollen and the flints are far from coeval. It will be helpful, however, to use such comparative dating of the artifact assemblages as can be made on the basis of their typology (Jacobi 1976; Shennan and Schadla Hall 1981).

The sampling pattern for pollen was the same in each of the cases described here. At a point selected by the excavator as being well within the apparent confines of the site, a contiguous series of 1-inch (2.5 cm) or 2-cm samples was taken from the subsoil (archaeologically sterile) right up to the present soil surface.

Iping, West Sussex

The first of the sites to be discussed here is Iping (Keef *et al.* 1965). The site was a spread of microliths lying on gently sloping ground on Iping Common. The whole area today is heathland, dominated by heather (*Calluna*), and the soil is a strongly leached heath podzol developed in a medium to coarse sand. The pollen diagram is shown in Figure 26. On the scale of this diagram the main flint concentration was at about 12 inches depth. The industry is of Maglemosian character, and therefore to be regarded as Early Mesolithic, dated between 6,800 and 8,000 B.C. (Jacobi 1976: Pitts and Jacobi 1979). In terms of pollen zones this would fall around the Preboreal/Early Boreal boundary. In Figure 26 two diagrams are given. The upper one covers the whole profile and in the lower one the lower part of the profile is graphed at a larger scale. Zone A does not concern us here; it is the raw humus of the present heathland soil, so the bulk of the pollen is recent. (This site was devastated by an intense heath fire in 1976, which destroyed not only the vegetation but also consumed most of the raw humus at the surface of the soil.)

In Zone B there is a characteristic heath soil pollen sequence, with progressive decrease of hazel and a parallel increase of heather. There is also an element of agricultural pollen possibly from prehistoric agriculture on the site—there are several round barrows nearby—or alternatively from farming on land adjacent to the heathland, as is occurring today.

Zone B is demarcated from zone C by a major disruption of most of the pollen curves. This junction is shown on a larger scale in the lower diagram. Zone C, the zone in which the artifacts occur, is dominated by two taxa, hazel and heather. The heather curve is typical of a contemporary taxon developing on an exposed surface; clearly the junction was once a soil surface. On the

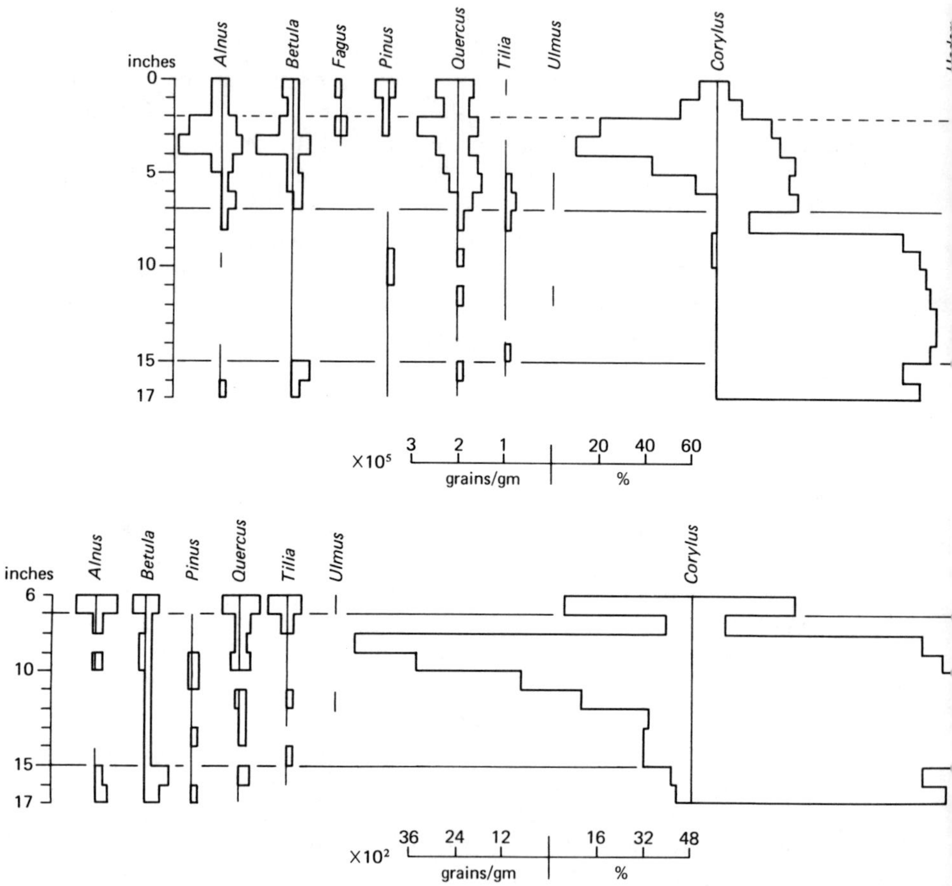

Figure 26. Pollen profile from Iping Common, Sussex. Upper diagram: complete profile. Lower diagram: sequence through a buried level at a larger scale.

other hand the hazel curve represents a taxon that has long been dominant but which has been ousted, in this case by the spread of heather. Such an ecological change has the appearance of an anthropological one, possibly through the use of fire. The pollen sequences suggest a change from dense hazel woodland to open heath, a change that would be accompanied by changes in the soil, as the pollen sequence shows. From 7 to 10 inches the hazel and heather curves show a definite pattern with depth, proving that no soil mixing was taking place. From 12 to 15 inches, however, the frequencies and percentages of hazel are almost constant. The boundary between zones C and D is not so clear-cut and may be difficult to substantiate. It has no

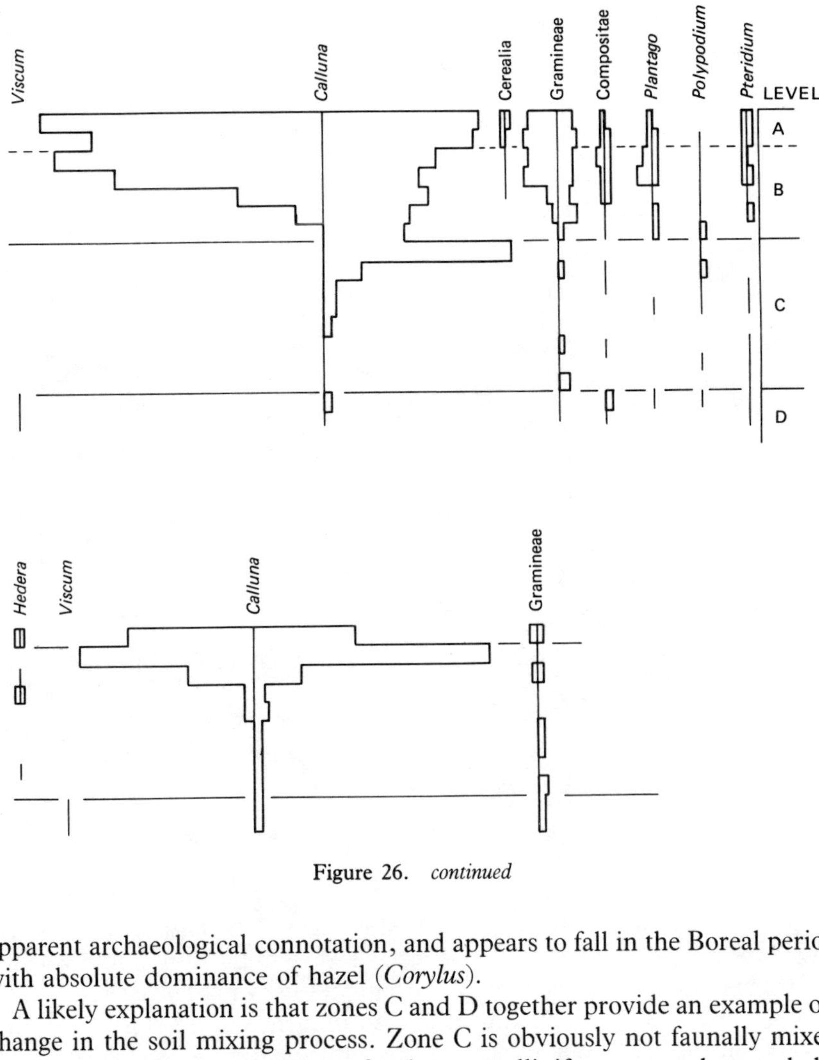

Figure 26. *continued*

apparent archaeological connotation, and appears to fall in the Boreal period, with absolute dominance of hazel (*Corylus*).

A likely explanation is that zones C and D together provide an example of a change in the soil mixing process. Zone C is obviously not faunally mixed, but zone D, though consisting of only two polliniferous samples, probably was. It could be maintained that zone D is the basal part of a homogenized soil (cf., Fig. 6) in the top part of which faunal mixing has ceased, allowing the development of the characteristic frequency distribution discussed in Chapter 1. Such a change need not be related to a sudden episode associated with man. Progressive acidification would eventually produce a pH which was inimical to the soil mixing fauna, but without bringing about a major change of vegetation. Eventually such a change happens too, as shown by the replacement of hazel by heather in the top part of zone C. Nevertheless,

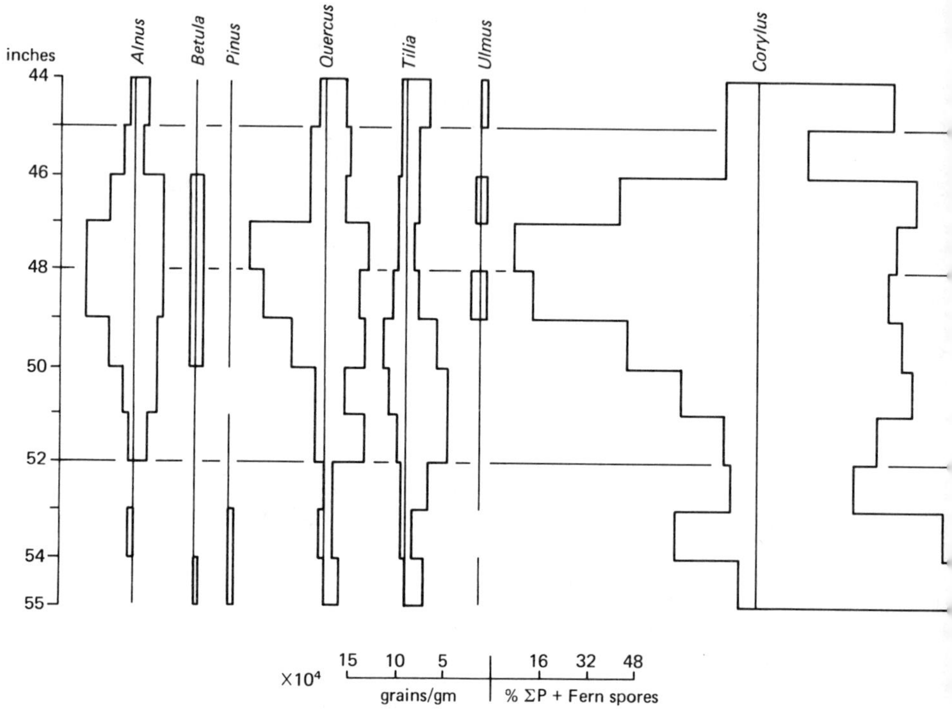

Figure 27. Pollen diagram through an old land surface and overlying turf in the core of a round barrow at Minsted, Sussex. (See also Fig. 15.)

though the earliest pollen phase here is dominated by hazel and is clearly Boreal, there is no proof that it can be correlated with the flint industry on the site. If the industry is indeed Early Mesolithic, it may predate even this early pollen record. However, continued human influence could still be exerting pressure on the ecology of the area even at this date.

Minsted, West Sussex

Minsted is not far from Iping and a number of mesolithic sites are known in the vicinity (Shennan and Schadla Hall 1981), but no pollen work has been done on them. However, when investigating the old land surface beneath a Bronze Age barrow (Drewett 1975), I found that the pollen record in the lower levels of the buried soil passed through a familiar compound profile (Fig. 27), the lower part clearly Boreal in date. There was also a high percent-

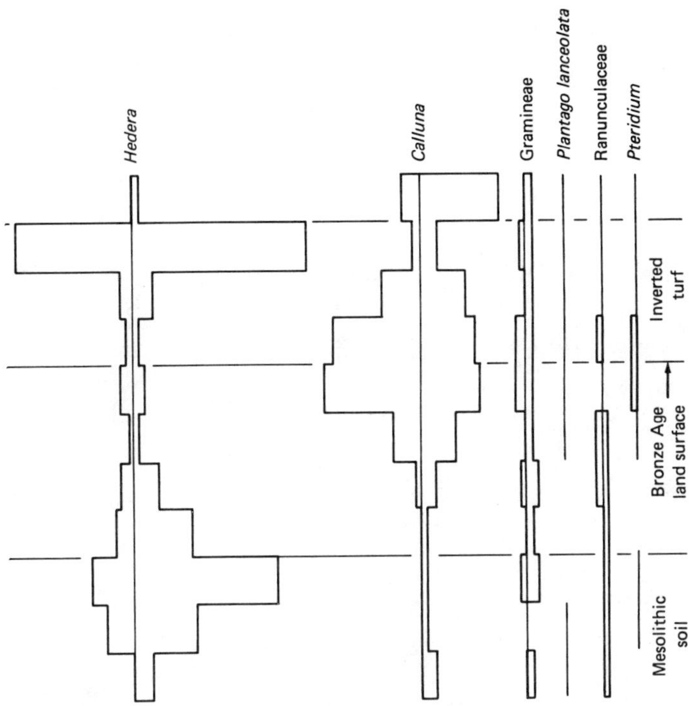

Figure 27. *continued*

age of ivy (*Hedera*) pollen in this level, very much as we shall see in some sites still to be discussed. Flints occurring beneath the barrow included a significant component of microliths, seemingly of a similar age to those at Iping. The upper part of the pollen diagram in Figure 27 shows an inverted turf at the base of the Bronze Age turf mound. This also has a high percentage of ivy pollen at its 'base', showing that the turf was cut from the old land surface, going deep enough to bite into the buried mesolithic level.

The Bronze Age land surface lies on a soil in which heather has become dominant. This could have happened at any time in the Neolithic, or even earlier, as the Boreal dominance of heather in the Iping profile bears witness. The spread of heather is to be seen as the result of man's activities and should not be regarded as a climatic indicator.

Oakhanger VII, Hampshire

The site of Oakhanger VII (Rankine *et al.* 1960) also provides ground for discussion of flint stratification. Rankine thought that there were three separate phases of stratification, but this is not reflected in the typology, and there

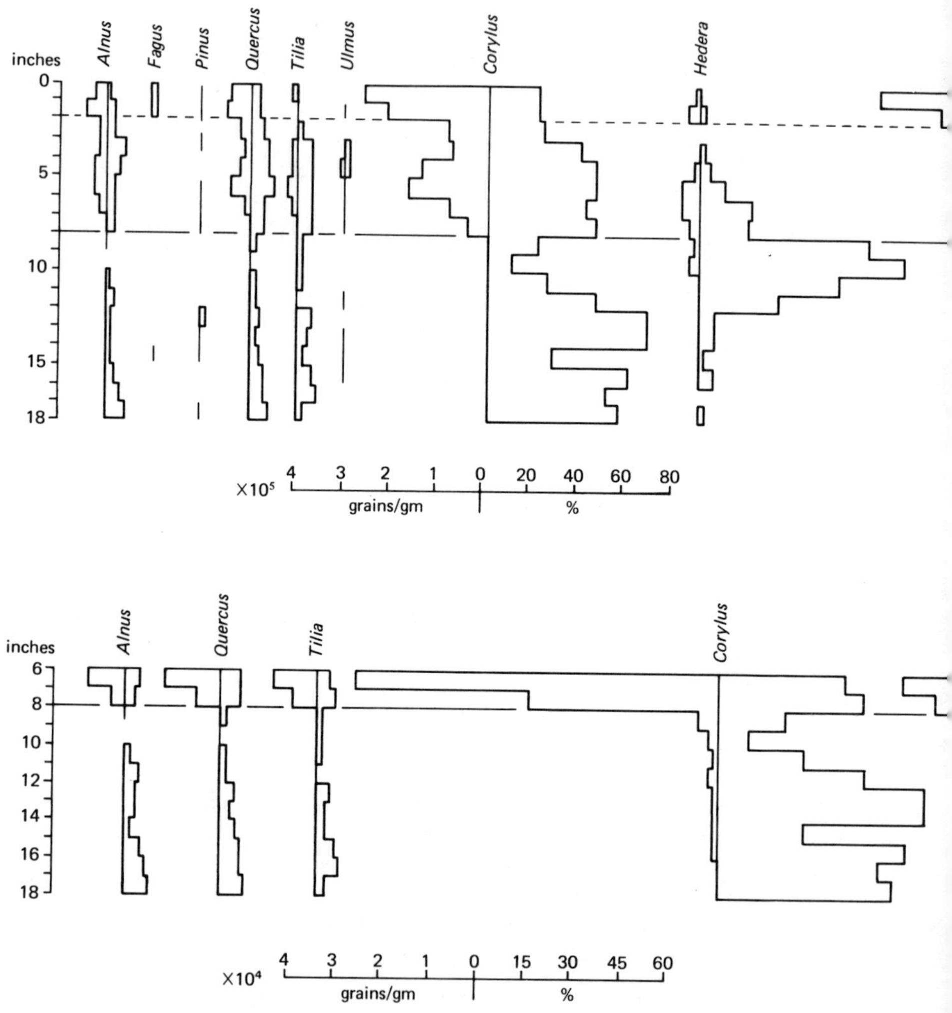

Figure 28. Pollen diagrams of the profile at Site VII, Oakhanger Warren, Hampshire. Upper diagram: complete profile. Lower diagram: a sequence through a buried surface at larger scale. C, Cerealia.

is some suggestion of a vertical scatter of flints between the so-called phases. Here too the industry can be regarded as Early Mesolithic, though perhaps later than at Iping, and probably falling in the early Boreal (Pollen Zone VI). A glance at Figure 28, however, makes it clear that in the very lowest levels of the pollen record hazel is dominant but pine, which should be present in

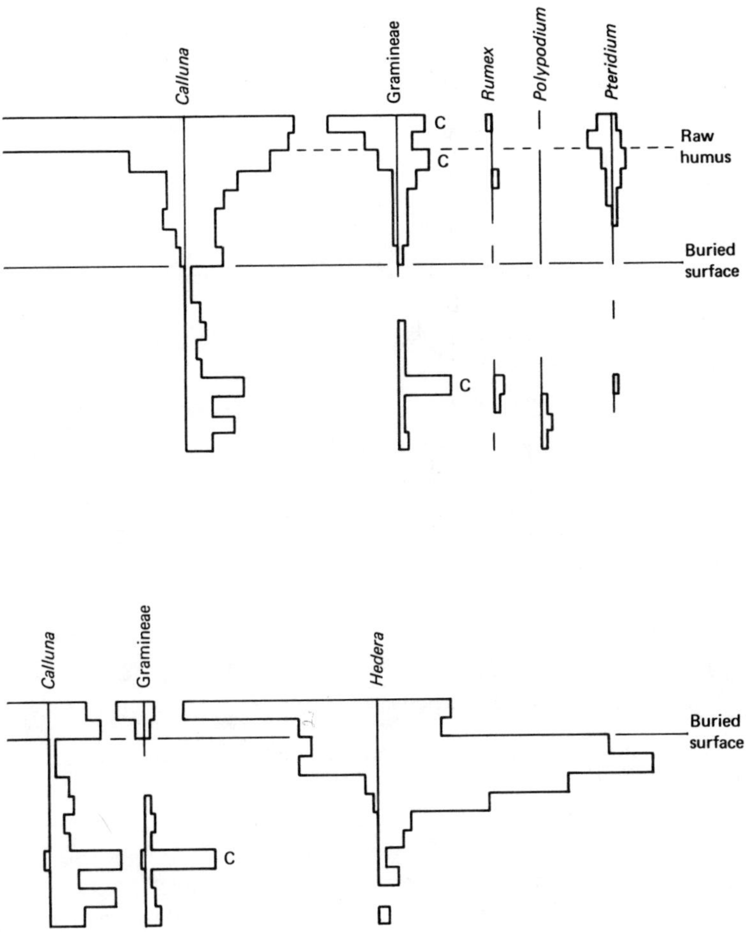

Figure 28. *continued*

Pollen Zone VI, is insignificantly represented. The consistent presence of alder, oak, and lime pollen in the 8–18 inch zone may indeed indicate that this pollen assemblage is Late Boreal if not Early Atlantic, yet the first industry is Early Mesolithic. It appears, therefore, that the pollen has not survived from the time of the archaeological occupation; at that time the pH of the soil may have been too high to permit preservation, and it was not until the Late Boreal at the earliest that it had dropped sufficiently for the pollen record to start. Indeed, the uniformity of the percentage curves in the 8–18 inch zone suggests that at this stage earthworm mixing was still taking place.

An obvious and remarkable feature of this zone is the massive representa-

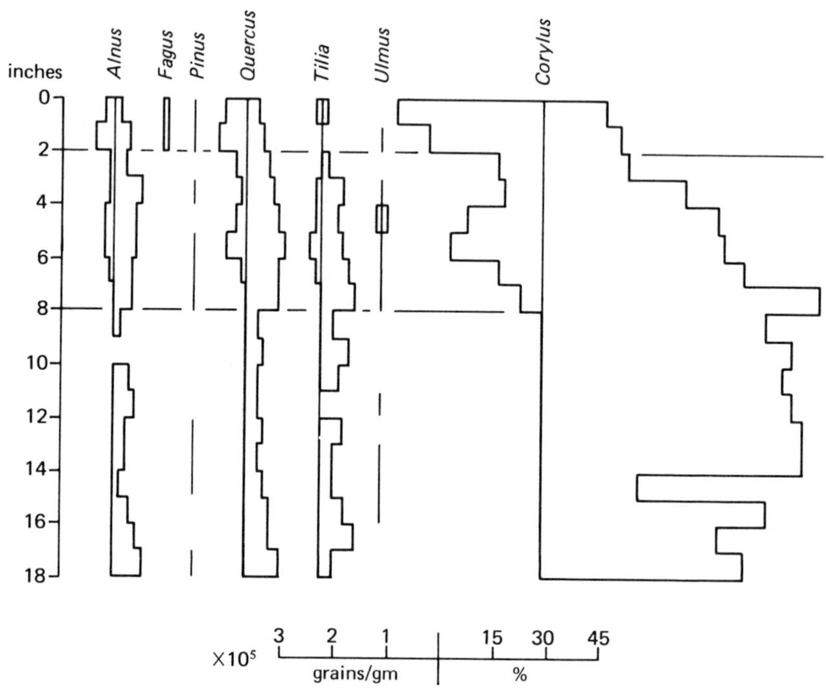

Figure 29. Pollen diagram of a profile at Site VII, Oakhanger Warren, Hampshire, excluding *Hedera*. (See also Fig. 28.) C, Cerealia.

tion of ivy (*Hedera*). It has been argued elsewhere (Simmons and Dimbleby 1974) that such accumulations—of which many examples are now known—may be an artifact, perhaps arising from the collection of the evergreen ivy as a winter fodder; it flowers in October and November. This need not imply the tending of domestic animals; mesolithic man was very dependent on the red deer, which is quite partial to ivy, and he may have been practising some form of animal husbandry. We could be seeing here evidence of a much later mesolithic activity than is represented by the artifacts on the site.

If the ivy component is intrusive, then it may be instructive to recalculate the data excluding ivy from the pollen total. When this is done Figure 29 is produced. The uniformity of the 8–18 inch zone is now clearer than ever.

There is a sharp break at 8 inches right across the diagram. Above that level pollen is much more abundant and the deciduous trees are more strongly represented. Hazel is dominant at the base but falls away and eventually heather, grasses, and bracken, a typical heath assemblage, take over. Ivy pollen, presumably derived from the underlying zone, also fades out, perhaps

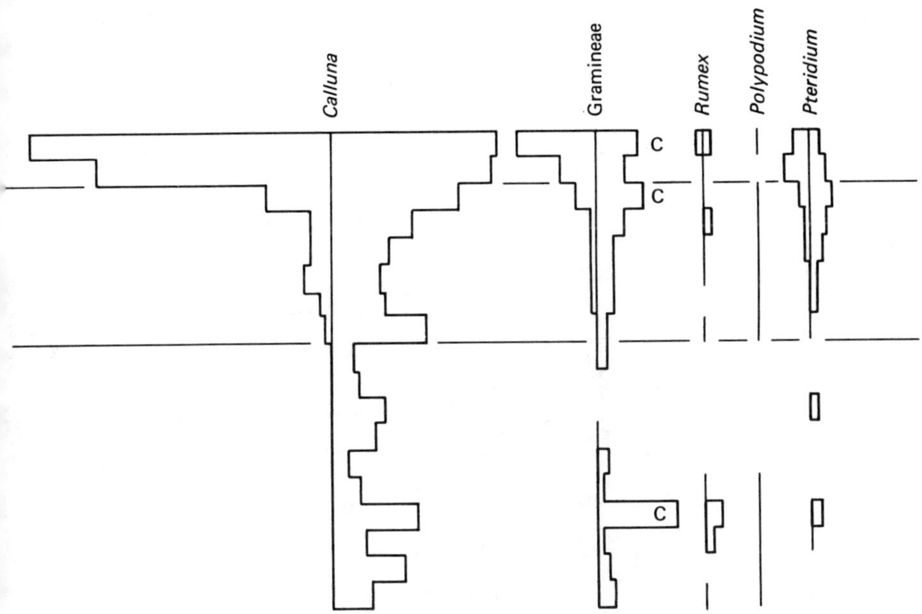

Figure 29. *continued*

to downwash, but it is interesting to see a trace of it in the present raw humus. This must be derived from some exposure of deeper layers, perhaps even as dust from Rankine's sondages.

The pollen in the 2–8 inch zone does nothing to date the material it is in. There is a scatter of microliths through this zone, and some even occur on the surface. This implies that the buried level became covered, perhaps by blown sand (see later) during the Mesolithic. Having become stabilized it then encapsulated a crude pollen record of the subsequent vegetation history.

Longmoor, Hampshire

Staying with the Wealden sites, the pollen profile obtained from Jacobi's site of Longmoor (Jacobi 1980, 1981) is shown in Figure 30. The flint industry, with the rare exception of one or two pieces, was Early Flandrian in date. The pollen record, however, shows a strong element of thermophilous trees even in the lowest levels; it is apparently no earlier than Atlantic in date. So again there is a long hiatus between the industry and the onset of the pollen record. It is, however, interesting to note that there is a considerable representation of heather pollen throughout the profile, which does not seem to be related to the surface dominance of heather, and also a liberal sprinkling of

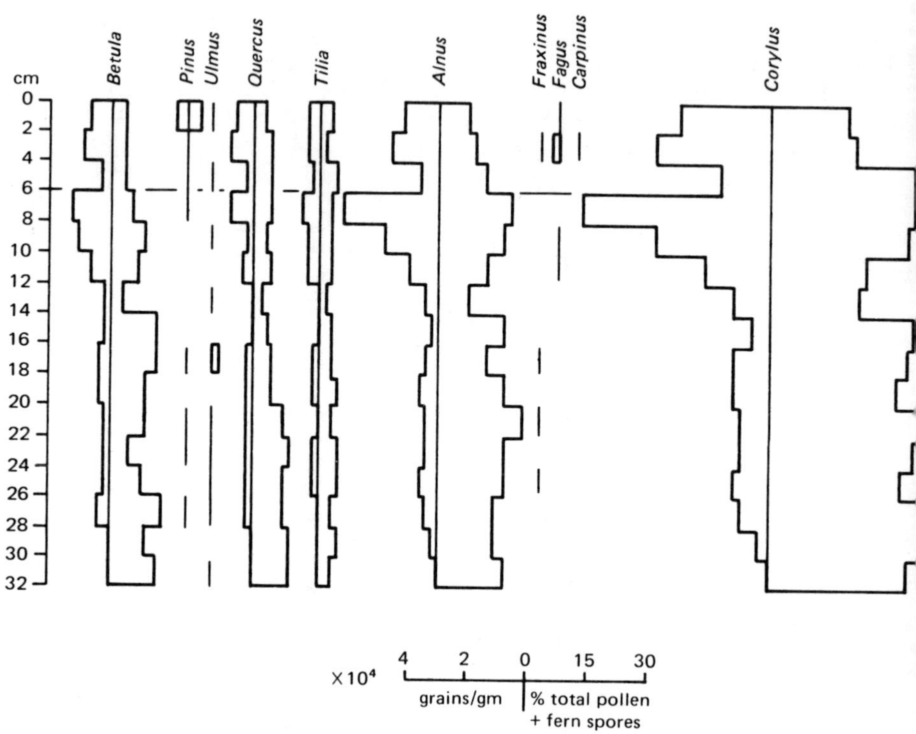

Figure 30. Pollen diagram of a soil at Longmoor, Hampshire.

herbs of various sorts. There is no cereal pollen, and it seems likely that what is reflected here is opening up of the forest, though not necessarily for agriculture.

In the Longmoor profile too there is a sudden break, but this time it is shallow, only 6 cm below the surface. However, the presence of pollen of deciduous trees above this level suggests that it is a truncated profile which has been exposed long enough to allow the accumulation of some heather and grass pollen, together with some recent tree pollens such as hornbeam (*Carpinus*) and beech (*Fagus*). There is no overrepresentation of ivy pollen in this analysis.

Winfrith, Dorset

From here it is a logical step to the Winfrith diagram (Fig. 31; Palmer and Dimbleby 1979). Here I only want to comment on the zones PI and PII,

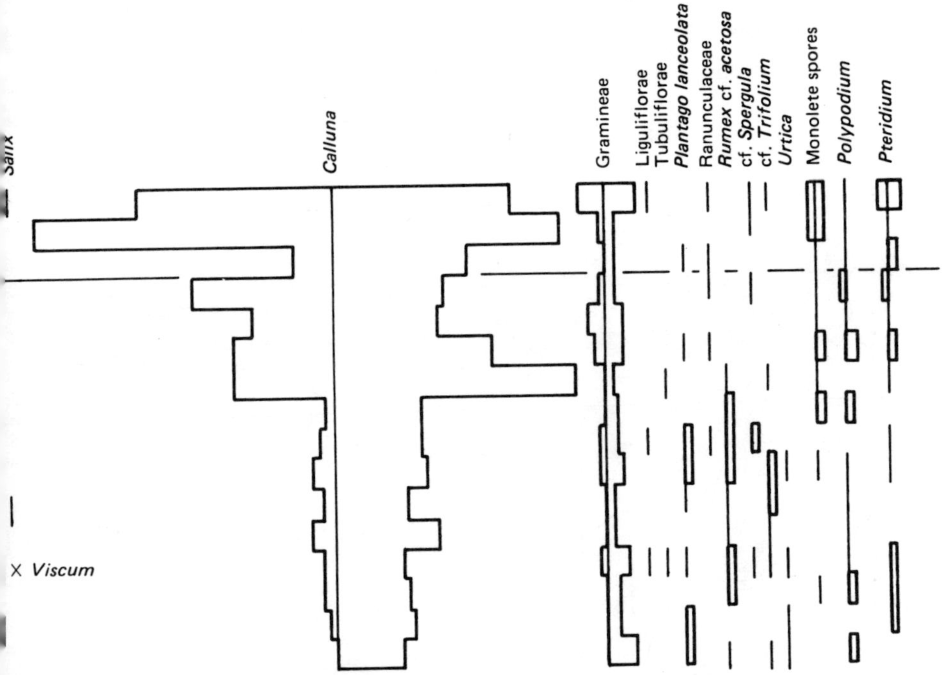

Figure 30. *continued*

which are sharply separated from each other. PII, the lower layer, is dominated by the pollen of deciduous trees and has a negligible component of the pollen of light-demanding species; the most abundant of these is bracken (*Pteridium*), which is a woodland plant, but only produces spores freely in better light conditions. In PII there is also a massive component of ivy pollen, and again, on the theory that this is introduced by man, the graphs have been recalculated excluding this taxon (Fig. 32). These curves suggest a period of more open tree canopy, as indicated by the bulges in the curves for *Calluna* and *Betula*, but surprisingly not *Pteridium*. The tree pollen is characteristic of the Atlantic period. There seems to be some doubt about the age of the flint industry; at one point Palmer describes it as Late Mesolithic, but elsewhere suggests it is of similar age to Oakhanger. Either way, there is no means of correlating the occupation with the pollen. Even the ivy episode, assuming it to be the result of mesolithic activity, cannot be correlated with the flints on the site. This episode certainly occurred in the Atlantic period, but it could have been brought about by a quite separate group of people.

The sharp break separating PI and PII continues the familiar pattern with the exception that here pollen frequencies are much higher in PII than in PI. Whilst there are traces of pollen from PII in the upper layer, PI, notably

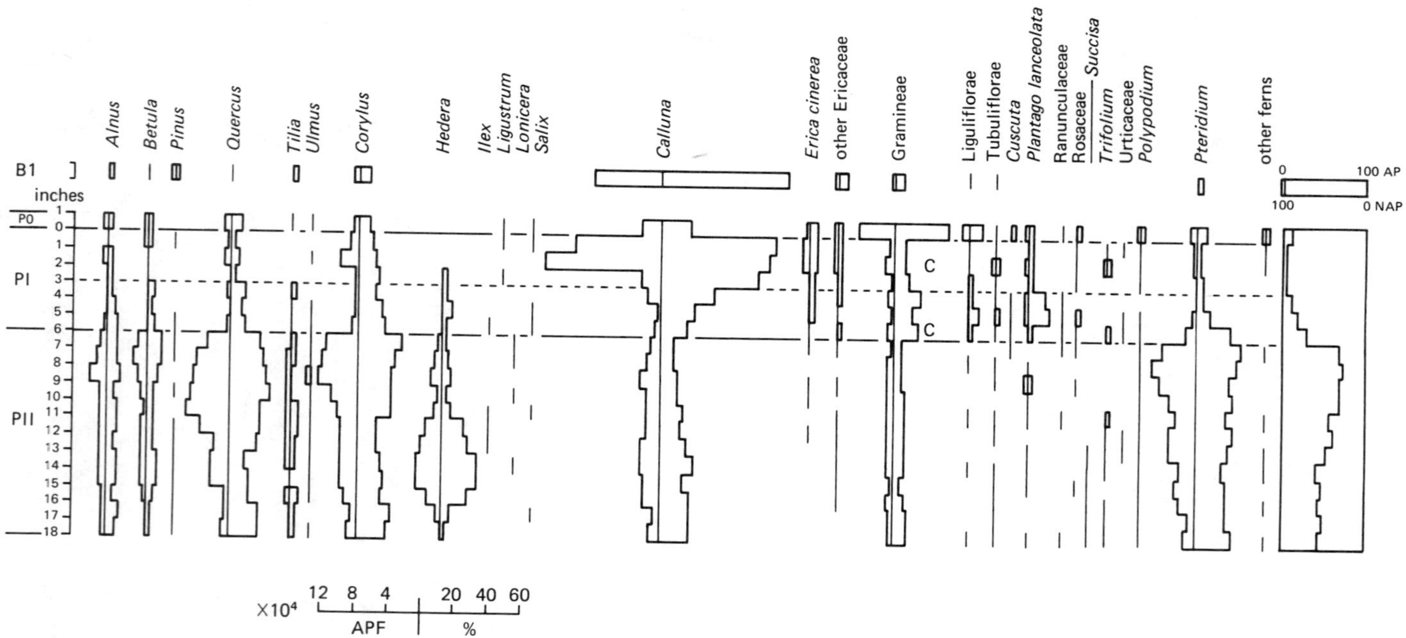

Figure 31. Pollen diagram from Winfrith Heath, Dorset. C, Cerealia.

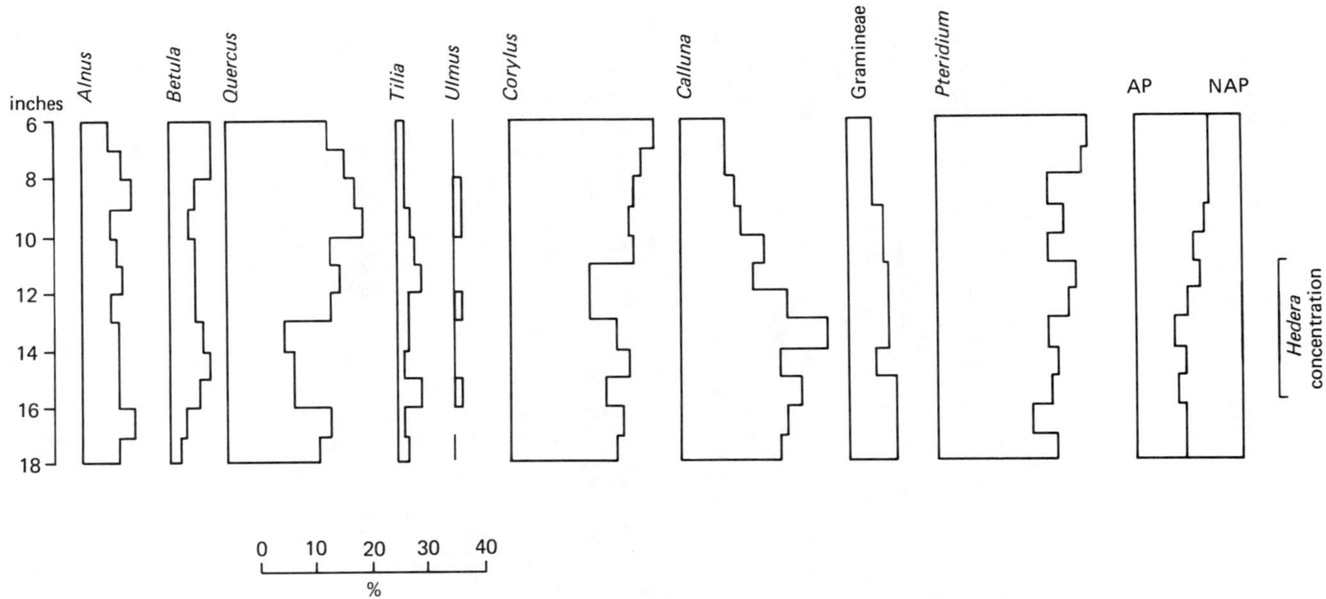

Figure 32. Pollen diagram from Winfrith Heath, Dorset, excluding *Hedera*. (See also Fig. 31.)

Hedera and *Tilia,* the bulk of the pollen here is post-agricultural, showing a very high NAP/AP ratio, and the ultimate dominance of *Calluna*. The main point of interest is the suggestion, as at Oakhanger, that this layer was laid down long enough ago to have collected a long pollen record, in this case ranging from the beginnings of agriculture onwards.

Addington, Kent

Finally, of the sites I have worked on, we come to Addington (Dimbleby 1963b), which is not in the Weald but on the Folkstone Sands in Kent. Though I am treating it as an open site it was in fact buried, beneath the recent up cast from a sandpit. In Figure 33 we need only consider zones C and D; A and B are part of the overburden.

Here again we have the usual compound profile. The lowest zone (D), where the main concentration of microliths occurred, is poor in pollen, but what there is includes mainly alder (*Alnus*) and oak (*Quercus*), with a minor element of hazel (*Corylus*). The grasses and light-demanding herbs are very poorly represented. Though not in large quantities, lime (*Tilia*) is present throughout, and this together with traces of elm (*Ulmus*), could point to an Atlantic date.

In the overlying zone C pollen is preserved in much larger quantities, still dominated by the deciduous trees, but these are progressively phased out and supplanted by grasses and bracken. The flint industry had been described as Late Mesolithic, so it is appropriate to look for evidence of anthropogenic influence on the forest composition. However, there is no episode to be seen in the pollen record which could match up with such an explanation in the lower part of the diagram. There seems to be no identifiable clearance until the upper part of zone C, but this is accompanied by cereal pollen and pollen of agricultural weeds. Possible explanations of this lack of apparent impact have been discussed elsewhere (Simmons and Tooley 1981) but one simple answer remains—namely that the whole pollen record postdates the occupation, even if this can be described as Late Mesolithic.

Montbani, Aisne, France

All the examples quoted occur in Southern England, but as pollen analysis has rarely been applied in this way elsewhere one cannot tell how widespread this phenomenon of compound soils might be. However, I have come across one example, from Montbani (Aisne, France), which has some parallels. The pollen analysis was carried out by Planchais (1972) on samples through an open mesolithic site on a sandy soil. As Planchais presents her data in a manner which does not make comparison with mine easy, I have recalculated

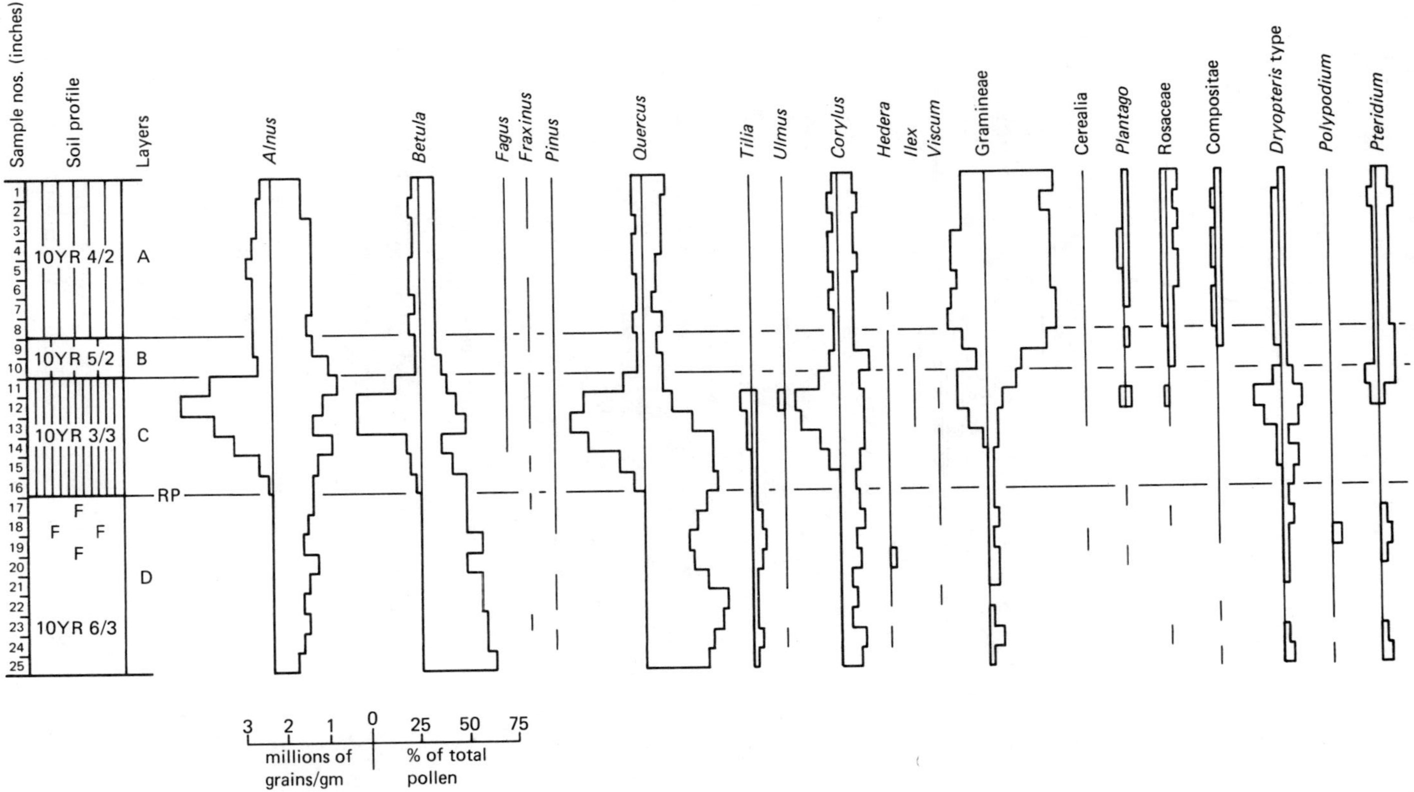

Figure 33. Pollen diagram through a soil profile, Addington, Kent. For soil profile, intensity of shading indicates darkness of humus colour, also indicated by Munsell colour notation. FF, flint zone in silver sand; RP, root pan.

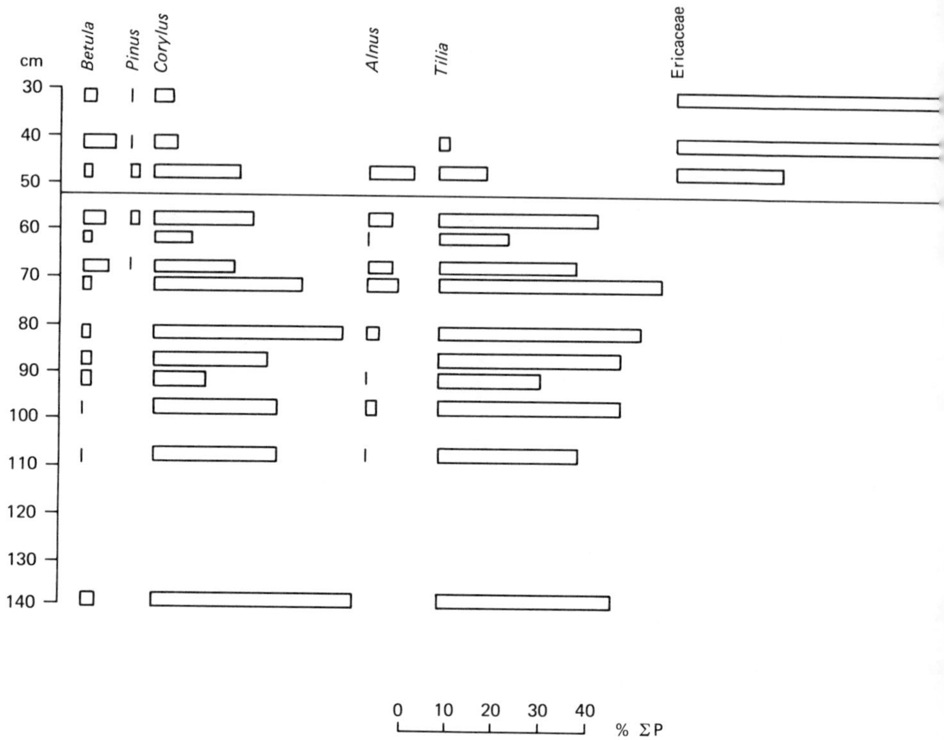

Figure 34. Montbani, France. The pollen diagram is based on recalculation of Planchais's (1972) data.

them from her diagram and presented them as Figure 34. Though there are only percentages (of total pollen) to work on, the familiar pattern of a compound soil seems to emerge. There is a break round about the 50 cm mark, below which the pollen is primarily of hazel and lime, but alder is also present and the suite appears to be of Atlantic age. Above 50 cm these taxa fade out, to be replaced by a massive element of Ericaceae.

However, in the lower zone there are other taxa which assume large proportions; one is ivy (*Hedera*) and the other is Asteraceae (= Compositae). We have already seen that the ivy pollen may have been artificially introduced into such a context, but what about the Asteraceae? I cannot suggest a convincing answer for this particular case, but in Chapter 7 a number of examples are given of overrepresentation of various groups of the Compositae. They are nearly all insect-pollinated plants, so these large quantities are plainly not in equilibrium with the pollen rain. But through animal food

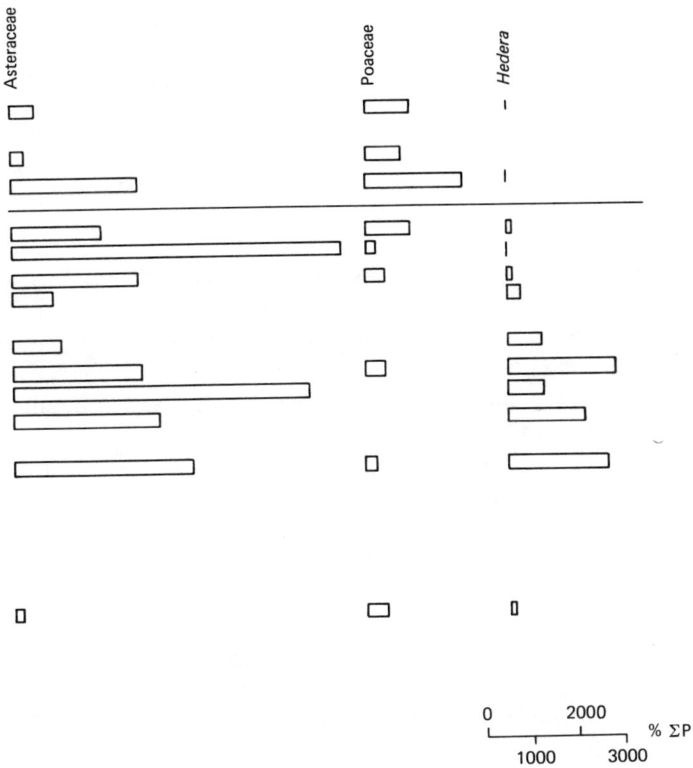

Figure 34. *continued*

chains and by other means it seems possible for these high concentrations to be produced, sometimes naturally, sometimes through human agency.

DISTRIBUTION OF ARTIFACTS

Though all these sites are being dealt with as open sites, because artifacts can be found on the present surface and there are no superimposed earthworks, it has to be recognized that by far the greatest number (or weight) of artifacts have to be extracted by excavation; they are buried in the sand. I do not think it is unfair to say that archaeologists have paid great attention to the microlith typology but have shown little curiosity about the fact that the flints are buried or how they become buried. On site there is sometimes speculation, but this rarely reaches the final report. Already we have seen that the pollen analyses provide some new facts, if not clear-cut explanations, and it is for this reason alone that some further discussion is appropriate here.

There are two features which call for consideration:

1. That artifacts are distributed vertically as well as horizontally, sometimes as much as 60 cm below the present surface (e.g., Oakhanger).
2. That the pollen reveals that all these profiles are compound, with an *in situ* soil overlain by a deposited one, commonly about 15 cm in depth, also containing mesolithic artifacts.

In the light of the second of these features we do not have to assume that the artifacts in the basal soil came from the present surface. They could have originated from the buried surface. In some cases some flints are as much as 40 cm below such a surface (e.g., Oakhanger and Winfrith); in others the distance is much less.

The pollen evidence characteristically shows deciduous forest in this basal phase, which would imply a brown soil with earthworms present. At Iping there is indication that the mixing fauna disappeared even before the basal soil became buried, but there is also slight evidence that the soil was originally 'homogenized'. Earthworms will bury objects to a fixed maximum depth, given enough time (cf. Rackham, above). In calcareous soils this is around 20 cm and it may take up to a century for all the objects to reach this level. I know of no comparable figures for acid soils, but one would expect the ultimate depth to be less. If earthworm activity ceases before all the objects have reached their full depth, some will be 'stranded' above the theoretical maximum depth, though the majority may have reached that maximum, producing an apparent concentration.

On the face of it earthworm activity seems inadequate to explain the burial of artifacts to depths greater than 20 cm; the pollen diagrams lend no support to the suggestion that even these basal soils contain more than one surface. However, it must be remembered that the pollen can only reveal such a surface when it is actually present in the soil; it is quite possible that there was a series of surfaces buried during the earlier mesolithic period at a time *before* pollen was being preserved. Indeed, at Iping pollen was preserved earlier than anywhere else, and it revealed both soil acidification and the eventual burial of a surface, still in the Boreal period.

Archaeologists often invoke trampling as a means of burial. Whilst this no doubt could account for shallow burial of objects, I question whether it could account for burial to the depths required here, except possibly for the isolated flint.

To turn to the second feature mentioned above, the compound soil profile revealed by the pollen analyses, it is apparent that this occurred long ago; mesolithic artifacts are found in and on the upper deposits. Shennan and

Schadla Hall (1981:17) describe a hollow 'which was gradually infilled, apparently during the Mesolithic, with over a metre of brown [*sic*: blown?] sand'. The pollen diagrams of most of the sites under discussion here are compatible with the burial of the basal level by sand, still in the Mesolithic period. It may be noticed that in those sites which have 'intrusive' pollens in large quantities, traces of these are found in the overlying deposit, confirming the origin, if not the actual date, of the superficial layers (*Hedera* at Oakhanger, Minsted, and Winfrith; *Hedera* and Asteraceae at Montbani).

It should be noted, however, that at Oakhanger the *frequencies* of *Hedera* reach higher values in the upper zone than they do in the lower one though the percentages are much lower. On the face of it this seems to indicate that the material of the upper zone could not have come from the basal soil. But there are some crucial unknowns; for one thing we know nothing about the lateral variation of *Hedera* pollen in the basal soil. At some points it could have reached much higher frequencies than at our sampling point. Nor do we know anything about the relative rates of pollen decomposition in the two layers; over the millennia differences here could completely change the picture.

Most of the upper layers reveal a consistent pollen sequence showing the decrease of the forest and the spread of agriculture. There is no evidence of soil disruption during the period. It may be remarked that once agriculture had become general heathland areas were of little interest, except as rough grazing.

The most obvious mechanism for the formation of these compound profiles is by wind-blow, possibly operating on an exposed or disturbed surface. The difficulty about this explanation is that artifacts, and even the occasional stone, may occur in the deposit. There are serious objections to postulating earthworm activity at the new (i.e., present-day) surface. It seems more probable that accumulation was spasmodic during the Mesolithic, allowing objects to be dropped and subsequently buried.

One can speculate on this theme, but until archaeologists show an interest in this whole question of vertical distribution of artifacts and its relation to soil instability, and deliberately collect precise field data, we shall not progress far beyond speculation. As we have seen, Jacobi is already making some progress in this direction. In the hope of stimulating interest I will end this chapter with my own speculation. These base-deficient sandy soils once carried deciduous forest, with a brown soil containing earthworms. Mesolithic activity and occupation disturbed the ecosystem, leading to a series of episodes of minor disturbance, followed by wind-blow, perhaps even before the earthworms had been driven out by increasing soil acidity. When this eventually happened the soil structure was destroyed, allowing more massive movement, but even at this stage there was probably more than one episode.

At some point during the course of this process the soil became acid enough for pollen to be preserved, which on analysis revealed just those events that took place from this time on. It could not illuminate events that took place before it began to accumulate, obviously, such as earlier phases of soil instability; the distribution of the artifacts at present is our only clue to such happenings. I suspect that the processes of earthworm activity, soil acidification, soil instability, and wind-blow are sufficient to explain the complexity of these open sites, but on this we must await further field research, and I may well be proved wrong.

These compound soils remind me some of the 'buried levels' discussed in Chapter 3, or of sites having no known archaeological connections. The phenomenon may be much wider than we realize, and one must wonder whether the influence of prehistoric man has been much wider than we have given him credit for.

6
CAVES AND ROCK SHELTERS

CAVES AND ROCK SHELTERS

Though often lumped together by archaeologists, perhaps because they had similar attractions as occupation sites for prehistoric man, caves and rock shelters are different both in origin and in their usefulness as sources of pollen evidence. Caves characteristically occur in areas of soluble rocks, where they are formed by ground water solution. Their layout and dimensions may depend on the joint structure of the solid rock, and they can run underground for many hundreds of kilometres, one system perhaps having several entrances. It is sometimes unbelievable that prehistoric man could have penetrated, as he did, to some of the remotest parts of cave systems; it is even more extraordinary that he should have wanted to.

By contrast rock shelters do not penetrate far into the solid geology. A simple overhang may provide sufficient shelter to have attracted prehistoric man; when on a really large scale massivc arched strata can produce an extensive favourable microenvironment. Other so-called shelters may be fashioned by the lateral tearing action of fast-flowing water or slow-moving ice. These conditions may have obtained in a very different climatic era, and it is not uncommon for shelters created by these means to be very remote from water or ice today; many so formed are now in desert areas. Other mechanisms include heat fracture of exposed rocks, freeze and thaw, and other weathering processes. Not being normally penetrated by streams and rarely by seepage, rock shelters do not characteristically contain water-laid deposits. Moreover, being dependent on mechanical processes rather than chemical solution for their formation, they do not accumulate solution residues. On the other hand they may accumulate very substantial deposits derived from mechanical weathering of their roofs and walls. To this may be added dust and other airborne detritus blown in from outside. In this they have a close similarity to cave mouths.

The deposits in caves, and to a less extent in rock shelters, have been a prolific source of palynological information about vegetation changes from Pleistocene times and up into the Postglacial. The deposits are frequently layered, consisting of rock debris falling from the roof, especially under the influence of frost in cold periods, and often intermixed with water-laid deposits. In many cases it has been shown that a chronological sequence of pollen spectra can be built up. In this sense, therefore, accumulating cave deposits serve a similar purpose to accumulating peat or lake sediments. There are important differences, however: it is rare, for instance, for a column of samples to be taken; usually samples are taken from separate layers whose chronological sequence can be established. One might therefore regard cave deposits as outside the scope of this book, were it not for the fact that caves were frequently occupied by man, particularly in the paleolithic and mesolithic periods, and consequently they may very definitely be called archaeological sites. Many caves which have no archaeological remains in them have been investigated palynologically and those that do contain human remains have been generally been investigated primarily as a source of stratified pollen, and not specifically as archaeological sites; the artifacts and associated features in such cases are an added bonus—in effect zone fossils which can give a dating fix. In other cases, however, the primary investigation has been by archaeologists, who have called in the pollen analyst to provide an environmental background for the cave culture, usually in order to set the culture in a climatic context which can then be used to put it in an approximate date bracket. Cave palynology has provided a climatic sequence for the last 40,000 years (Leroi-Gourhan, 1973) and some sites have yielded pollen evidence even earlier than this. For example, Girard and Renault-Miskovsky (1979) found a sequence going back to the later part of the Riss glacial at Coupe-Gorge. Such sites may be beyond the range of radiocarbon dating, so charcoal, even if it is present, is useless for dating.

Cave deposits have assumed a particular importance because they occur in regions where other sources of pollen evidence are scarce. By their very nature they occur in areas of soluble rocks, mainly limestones, areas which are deficient in peat bogs or lake sediments, and of course in which the soils themselves are for the most part calcareous and therefore of little use for pollen analysis. Thus a great deal of work has been done in France (particularly in the southwest), Spain, Greece, and other limestone regions in the Mediterranean basin. Although pollen analysis has been applied to cave systems in North America, it has so far not produced the same corpus of knowledge of past climates that has resulted from work in Europe. As we shall see, attention in America has been more specifically directed to the archaeological remains of man himself, and has not laid the same emphasis on long-term climatic change. By comparison with Europe, of course, the history of man in America is relatively recent, only from the end of the last glacial onwards.

Nevertheless, many pollen analyses of cave and rock shelter deposits have been made in the United States, frequently with the aim of establishing the plant environment. Reference has already been made (p. 26) to the work of Hall (1981) on the bias in pollen spectra brought about by deteriorated and damaged pollen, and it is significant that a number of his examples of such conditions come from caves and rock shelters, where the deposits are more-or-less aerated.

Much of the pioneer work on the pollen analysis of cave deposits was carried out by Mme. Arlette Leroi-Gourhan, the wife of archaeologist André Leroi-Gourhan, so it is no surprise that her botanical work has been closely integrated with archaeological research. However, her approach has been almost entirely that described above, namely the placing of archaeological horizons in their climatic setting. This is epitomized by her very comprehensive work at Lascaux (Leroi-Gourhan and Allain 1979) where she pieced together analyses from cross-dated levels throughout the cave system to provide a continuous and consistent series of pollen spectra through the so-called Lascaux interstadial, about 17,000 years ago. Although evidence of human influence in the pollen assemblage was found (see below) this was not a palynological investigation of an occupation site for its own sake, in the sense in which this has been described in earlier chapters.

POLLEN DISTRIBUTION IN CAVES

As mentioned above, cave deposits may be well stratified and so provide a clear-cut chronological sequence. A good example of such a case is the stratigraphy at the cave of Coupe-Gorge, Haute-Garonne, investigated by Girard and Renault-Miskovsky (1979). On the other hand, deposits may show a disturbed stratigraphy due to water-sorting, gravitational slip, deposition in different parts of a cave system, and so forth, so that a simple chronology cannot easily be established. However, even in such circumstances it may be possible, by working out the horizontal relationships of deposits, to make up a chronological sequence of deposits from which samples can be taken representing a chronosequence. This is what Leroi-Gourhan did so effectively at Lascaux.

Given that by one means or another a sequence may be built up, the question then arises as to how meaningful pollen analyses from such deposits will be. What are the sources of pollen that finds its way into cave deposits?

In her earlier work Leroi-Gourhan, together with van Campo, investigated how modern pollen reached different parts of a cave system (van Campo and Leroi-Gourhan 1956). They did this by exposing slides covered with glycerine jelly at critical points in several caves in the same locality. These caves varied in their lengths, cross-sections, and whether or not there was a through draught of air. By these means they showed that pollen coming in by the cave entrance could normally be traced up to about 10 m from the entrance, but that this distance could be greater if the cave had a through draught. It was also shown that in general the pollen amounts in the deposits were very low, a point returned to below, but that where there had been human occupation the pollen concentration was much higher. It was considered that this was due to pollen being brought in on human hands and feet, and perhaps animal paws and fur. It was observed that the pollen of such situations was not primarily that of anemophilous taxa but was largely that of herbs, especially aquatics and plants of the water's edge. It was suggested that ferns could have been brought in for bedding and other plants as food. At Lascaux Leroi-Gourhan had recorded in one sample very high counts of *Salix* and had concluded that this arose from a small ephemeral lake near the entrance, though in this case she does not attribute the concentration of *Salix* pollen to human agency. At Lascaux, however, remains of complete inflorescences of grasses and of *Artemisia* were found, and anthers or clusters of the pollen of these taxa were extracted from the soil. Clusters of pollen grains could represent the presence of anthers or fallen flowers, and it is interesting to see this confirmed by the Lascaux evidence (see Ch. 7).

Van Campo and Leroi-Gourhan (1956) also observed that pollen could be brought into cave deposits by flowing water or by seepage, and this source of pollen has been dramatically illustrated by several workers who have found pollen in stalagmitic material in caves. Damblon (1974), working on stalagmitic deposits in a cave in Belgium, showed that the pollen in this material consisted mostly of grass and hazel pollen and was clearly an intrusion from the latter half of the Atlantic period. Even more striking were the results obtained by Bastin (1979), also working in Belgium, who systematically sampled stalagmites from top to bottom. Not only did he show that a stalagmite could quite dramatically record changes in the vegetation, but also that the pollen assemblages had a strong local flavour rather than a regional one. Bastin also concluded that stalagmitic growth could take place in interglacial, interstadial, and even full glacial climates. It is certainly a clear indication that pollen is being transported by water into the cave system. Had there been no stalagmite formation the pollen would have found its way into other deposits through which the water seeped. Most of the stalagmites investigated by Bastin were within 50 m of the cave entrance, but one was 600 m from the entrance and still yielded pollen. At Lascaux stalagmitic material was the sole source of postglacial pollen.

The observation that pollen entering the cave by the cave mouth can penetrate 10 m or more into the cave means that the pollen from rock shelters, as one might expect, will be strongly biased by, if not totally due to, the vegetation local to the cave mouth. If one is using palynology to determine the regional vegetation, particularly the general forest cover, this means that a rock shelter is not a reliable source of such information. On the other hand, the pollen spectra may be an interesting source of evidence about the plant cover immediately around the entrance and may be a source of information about the impact of the cave users on the immediate environs.

There is very little recorded discussion about the possible movement of pollen within mineral cave deposits. Pollen analysts have found that they can extract pollen and once the layer sequence is established they can produce a pollen diagram. Usually diagrams are based on spaced and not contiguous sampling. At Coupe-Gorge, for example, the spacing in a simple stratigraphy was about 10 cm but the authors claim that the consistency of the results fully justifies this.

Frequently sampling is done in close relationship to the stratigraphy: each recognizable layer is sampled, often only one sample per layer. In the light of experience, Leroi-Gourhan (1960) came to recommend more detailed sampling, e.g., at 5 cm intervals. Whether or not the process of downwash and differential decay, with which we are familiar in soils, also applies to cave deposits is impossible to say. Leroi-Gourhan (1956) did observe that two samples taken from the same stratum often give different results, and that the lower one tends to be richer in herbaceous pollen. This might suggest that there is a more-or-less consistent change of spectra with depth. On the other hand so many of the published sequences do show an apparent consistency that if there is any movement of pollen in the deposit it does not seem to be great enough to vitiate the method. Nevertheless, one would like to see some critical investigation. Some cave deposits are extremely coarse, others fine-grained, and generally all are poorer in humus than soils exposed to the atmosphere. All these are factors which might affect the stability of pollen in the deposits. Van Campo and Leroi-Gourhan (1956) themselves make the point that pollen analyses of cave deposits cannot be interpreted in the same way as those of open sites; the problem is very complex and demands parallel work in sedimentology, analysis of animal remains, etcetera.

POLLEN FREQUENCY

There are conditions in caves which are conducive to pollen preservation. Deposits which are anaerobic, even if of high pH and often showing strong manganese deposition, can yield high frequencies, and it has already been mentioned that pollen may be concentrated where there has been human occupation, particularly, apparently, where the occupation has been in narrow corridors in the cave (van Campo and Leroi-Gourhan 1956).

More characteristically, however, pollen frequencies in cave deposits are low, as one would expect in such an environment. Caves as opposed to rock shelters, which may be formed by other processes, are usually the result of rock solution in limestones and the accumulated deposits tend to be calcareous, often secondarily calcified. Except where there is local waterlogging they also tend to be aerated, sometimes excessively so in some of the coarser cave detritus. As a result, pollen may be so scarce as to be difficult to recover by standard extraction methods. However, techniques are now available, involving separation of pollen in heavy liquids, which make it possible to process larger amounts of material. Girard (Girard and Renault-Miskovsky 1969) has devised a modification of the Frenzel method which has proved very successful. At the Abri de la Cure in Switzerland standard methods had given completely negative results, but when reinvestigated by Leroi-Gourhan and Girard using Girard's method, a comprehensive diagram was obtained (Leroi-Gourhan and Girard 1971). At Lascaux, too, early attempts to extract pollen using hydrochloric and hydrofluoric acid were ineffective, but again the new method was successful in producing very consistent results (Leroi-Gourhan and Allain 1979).

As pointed out in Chapter 1 the reliability of these very low concentrations of pollen may be open to question. Do the spectra faithfully reflect the contemporary pollen rain, or is one merely concentrating resistant types or even extraneous grains of casual occurrence? For cave deposits, at least, the answer seems to be clearly indicated by the diagrams, which show a remarkable consistency from sample to sample and moreover contain long lists of species—not a characteristic of depauperated assemblages. In fact it is the good representation of herbaceous pollen, much of which we know is susceptible to microbiological attack, which makes this line of research so valuable. As Leroi-Gourhan (1973) explains, the traditional use of tree pollen as a basis for postglacial pollen diagrams is not applicable to cave deposits for they frequently cover major climatic changes, including cold periods in which trees were scarce and herbs dominant in the local landscape. The ratio of arboreal to nonarboreal pollen is therefore more valuable in such a context and it is crucial that the herbs such as grasses and Compositae should be reliably represented. As we shall see below, they sometimes appear to be overrepresented, but that may have other explanations.

Even where adequate pollen counts cannot be obtained the interpretation of a spectrum may be in no doubt. For instance Butzer (1971) refers to pollen work in the Sahara, mentioning a sample from one rock shelter which gave a count of only 87 grains; however, 56 of these were of tree species, their presence being substantiated by macro remains. Leroi-Gourhan (1956) quotes another example where the climatic setting of one cave horizon had been based on the animal remains in it; a subsequent pollen analysis only produced 28 grains but their identity made it necessary to accept a quite different climatic interpretation from that based solely on the faunal evidence.

SOURCES OF ERROR

Associated with the actual frequency of the pollen in the deposit is another characteristic that may be useful, namely the state of preservation. In Chapter 1 it was mentioned that the pollen in calcareous soils, though at a low frequency, was often remarkably well-preserved, giving rise to the suggestion that it had not been in the soil for a long time and therefore might reflect the contemporary pollen rain. A similar distinction was made by Horowitz in relation to the pollen in a cave deposit at Sterkfontein, S. Africa (Vita-Finzi, 1978:107). In the Middle Breccia, which contained remains of *Homo,* the pollen could be segregated into that which was in good condition and that which was badly preserved. His interpretation was that the latter was coeval with the deposit, whereas the better-preserved pollen was intrusive, probably introduced by percolating water. The fact that the well-preserved pollen indicated a warmer element of vegetation than the poorly preserved pollen confirmed that the two were not reflecting the same climatic phase.

In view of what was said earlier about the ways that pollen can reach cave deposits it is perhaps surprising that meaningful results can be obtained at all, but obviously that suspicion is not well-founded. Nevertheless, Leroi-Gorhan draws attention to potential sources of error such as fissures, cave rivers, mixing of levels, etcetera (van Campo and Leroi-Gourhan 1956; Leroi-Gourhan 1965). Some errors can be reduced by careful field work, rejecting any stratification that shows apparent disruption or a potential source of contamination. One of the imponderables is the reworking of earlier deposits; in some cases this may be quite easy to recognize by the presence of taxa obviously at variance with the rest of the pollen. An example of this, not only from cave deposits, is the presence of Tertiary pollen and spores, proving contamination with earlier material.

A somewhat different example of contamination was demonstrated by Welten (Bandi *et al.* 1954) in the Brügglihöhle, a cave system in Switzerland. As part of a comprehensive programme of scientific investigation, Welten carried out pollen analyses down two sampling lines within a few metres of each other. In both flint remains from a Magdalenian occupation were found in a deposit which contained pollen of Older Dryas age, in a sequence extending from the Late-glacial up through the Postglacial. However, the two profiles differed markedly. One contained a massive superimposition of pollen from thermophilous trees, with a humic forest layer at the top, whilst the other had the merest trace of such an element. Yet the two profiles were quite close together. The difference apparently arose from a roof collapse above the first profile, precipitating later pollen from the soil above onto the deposits in the cave, and allowing a forest pollen rain to enter the cave. In the other profile there was no 'window' above, and it remained protected from this vertical influence.

Two other points of interest were recorded by Welten. The secondary pollen, though younger, was more corroded and was darker in colour (a feature more usually associated with older pollen; see Ch. 1). He also observed that the Compositae pollen was more corroded than that of the Gramineae, which he regarded as primary. He suggested that this could be accounted for by the Compositae pollen having originated from animal excrement.

INTERPRETATION OF ANALYSES

As indicated earlier a primary reason for carrying out pollen analyses on cave deposits has been to establish a climatic sequence into which the culture levels could be fitted. This provided a form of dating of these levels, but also added to the evidence of the sequence of events through the Pleistocene, which was an interest far beyond the purely archaeological. Very little use has been made of this material to illuminate the contemporary human environment and the way of life that might have been practised. In so far as this has been done it has been based on the naure of the artifacts and upon the animal remains which are associated with the cultural levels, and only to a minor extent upon the pollen data. There are, of course, exceptions to this general assessment, such as the appearance of cereal pollen in mesolithic levels (Leroi-Gourhan and Girard 1971; Renault-Miskovsky 1971). Such interpretation as there has been in terms of landscape has been based primarily upon the NAP/AP ratios, together with conclusions drawn from plants regarded as indicators of climatic or other ecological factors. If one tries to envisage these interpretations in terms of recognized plant communities considerable problems arise. For instance, the analyses of the sediments of the cave of Isturitz (Basses-Pyrénées) show a chronological sequence in the nonarboreal pollen from a grass-dominated assemblage to a long series dominated by Compositae-Cichoriae, to the virtual exclusion of anything else. In her interpretation Leroi-Gourhan (1960) refers to the invasion of the Compositae. I am unaware of any plant community with this floristic structure, or a succession that could produce it, though it has to be admitted that communities may have existed in the past that are no longer extant today. It seems more likely, however, that such an assemblage of pollen has been distorted and is not to be seen as a faithful representation of the landscape outside the cave.

This apparent overrepresentation of the Compositae (Cichoriae, Anthemidae, Carduinae, and sometimes other entomophilous taxa) is a widespread feature of cave analyses of all ages from Europe and the Mediterranean basin. Insect-pollinated plants do not produce large amounts of pollen, nor is it widely dispersed, so these are not plants whose pollen one would expect to find penetrating the remoter parts of caves; they often far outnumber the pollen grains of grasses, which are, of course, wind-carried. There is one other consideration that might be mentioned here. The occurrence of excessive amounts of pollen of *Hedera* (ivy), another entomophilous plant, has already been referred to and it has been suggested that it has been concentrated by man, perhaps through collecting it as green fodder. Could it be that the high percentages of Compositae pollen are also due to a bias introduced by man? We have already drawn attention to the fact that entire plants of a few species (e.g. *Artemisia,* which is in fact wind-pollinated) have been found in caves, suggesting that man brought them in. However, there is no such record for the plants of the three subfamilies mentioned above. The hypothesis that man carried the plant remains in is rendered even less tenable by the fact that high values of these pollens can be found in archaeologically sterile deposits; e.g., Cichoriae and Anthemidae in the sterile preboreal layers of the Abri de la Cure (Leroi-Gourhan and Girard 1971) and Anthemideae and *Artemisia* in the final Riss of the Coupe-Gorge cave, Montmaurin (Girard and Renault-Miskovsky 1979). As it is possible that spectra can be in some sort of equilibrium with the vegetation outside the cave one must consider what other factors may be at work, or may have been at work at some period in the Pleistocene. Van Campo and Leroi-Gourhan (1956), as pointed out earlier, believed that pollen could be brought into caves on the paws and in the fur of animals that either used the cave or were brought in by man. They suggest that the bear may be important in this respect. However, another animal that might be important here is the bat. It is a cave-dwelling animal and it is insectivorous, so that it may be feeding on insects which are visiting entomophilous flowers. In fact bats have been observed hunting over *Salix* catkins and catching the moths that visited the catkins (Harrison Matthews 1952:91). In such a way the pollen of entomophilous plants could be introduced into the bats' alimentary canals and voided within the cave. It will be remembered that Welten, in his work on the Brügglihöhle quoted above, had come to the conclusion that the corroded state of some Compositae pollen might be accounted for by their having passed through the gut of animals. Some bats will bring insects into their sleeping place and consume them there. Archaeologists do not commonly record the wildlife of the caves they are working in, so the presence of bats is not recorded. It also needs to be established whether bats were found during the colder climatic periods when the Compositae flowers were blooming. Many bats go into deep hibernation during the cold season, but during the summer they feed voraciously to build up a fat reserve.

This explanation could also account for the presence of other insect-pollinated plants whose pollen sometimes seems to be overrepresented in cave deposits. It could be an alternative explanation of the remarkable preponderance of *Hedera* pollen in the mesolithic levels in the Abri de la Cure at Baulmes (Leroi-Gourhan and Girard 1971) or the high values of Caryophyllaceae that are found in a number of cave analyses. On the other hand the high values of fern spores cannot be explained this way, and they are to be found in layers both with and without evidence of occupation (see again Baulmes). They are by no means consistently present in high proportions, but one has to bear in mind that where total frequencies are low there is the possibility of differential preservation, and fern spores are notably resistant to decay.

Apart from bats other cave denizens may have contributed to the pollen by way of their droppings. Pollen being remarkably resistant to digestion (Dimbleby 1978a:114) may be passed along a food chain and eventually voided by the carnivore at the end of that chain. Certainly pollen has been found in faecal deposits, both animal and human, from caves in North America, though in these cases the likelihood is that the pollen reached the faeces directly from ingestion of plant material. Coprolites, that is, subfossil faeces, will be discussed in a wider context in Chapter 7.

7
BIASED POLLEN SPECTRA

INTRODUCTION

As soon as one applies pollen analysis to archaeological sites, where by definition man has been at work, the possibility may arise that pollen spectra are not in simple equilibrium with the plant cover of the region but are biased in one way or another. Reference has already been made in Chapters 5 and 6 to the overrepresentation of ivy (*Hedera*) pollen, possibly through the collection of this plant for fodder, and this case will be returned to presently. It is perfectly possible, however, that a pollen assemblage should be biased even through perfectly natural causes. To take a simple illustration, the analysis of the pollen on a honey bee's leg after a foraging flight would bear no obvious relationship to the general vegetation, and as such would be dismissed for the usual purposes of conventional pollen analysis; on the other hand, if an apiarist should wish to find out what plants his bees visit, a pollen analysis of a bee's leg would be very informative (Hodges 1974).

SELECTIVE CONCENTRATION OF POLLEN BY ANIMALS

To begin with we will consider the natural action of some animals in concentrating certain pollen types selectively; the example of the honey bee has already been given. In actuality, the concentration of pollen in the nests of ground-nesting bees has been reported, a fact which is of course relevant as a possible source of contamination in soil pollen analyses. Faegri (1962) has examined the nest of a bumblebee, and Bottema (1966, 1975) has attributed high concentrations of Liguliflorae pollen to digger bees and mason bees. Possibly other burrowing insects such as ants may differentially collect pollen in their underground workings by accumulating pollen-rich organic material. The demonstration by Armitage (1975) that opal phytoliths from grasses could be extracted from the teeth of cattle suggested to me that pollen might similarly become attached to the teeth of herbivores. At that time I had in the laboratory a number of jaws of red deer recently culled from Scottish herds, so preparations were made from the bases of the teeth of several specimens. Pollen was indeed present, and interestingly the pollen assemblages varied, some being predominantly of grass pollen, others having a noticeably higher proportion of ericaceous pollen. I have never pursued this enquiry and have not applied it to old material, but, as with the phytoliths, it suggests a possible source of information about dietary habits of wild or domestic herbivores.

Dietary habits are also revealed by the pollen contained in the faeces of animals; in many cases fragments of plant foods are also extractable and identifiable. Subfossilized faeces, called coprolites, are not an uncommon feature of archaeological sites, especially in caves, as already noted (Ch. 6) and it is not only in the archaeological context that coprolites have provided valuable information. Laudermilk and Munz (1938) examined the fossil faeces of sloths of Pleistocene age to establish the feeding habits of these extinct animals. Palynology of similar coprolites has also been carried out with the aim of reconstructing the environment in which the animals lived (Martin *et al.* 1961).

On archaeological sites it is coprolites of human origin that occur most commonly, though I did once investigate coprolites, obviously of animal origin, found on an Iron Age site at Ivinghoe Beacon, Buckinghamshire (in Cotton and Frere 1968). The coprolites were quite rich in pollen, giving a different spectrum from and longer list of taxa than the surrounding soil. Surprisingly, and uniquely in my experience the predominant pollen type was from the family Rosaceae, mainly *Rosa canina* type and *Rubus fruticosus* type, neither renowned as abundant pollen producers. This raised questions about the animals' feeding habits; was it possible that the pollen came from a secondary source, such as the hive of ground-nesting bees? An animal that springs to mind in this connection is the badger, and though badgers will defecate in prepared places on open downland, it was zoological opinion that these were not badger droppings but most probably those of dogs. So far we have found no convincing explanation of why a dog should be ingesting dog rose pollen! It may be an example of a point made in Chapter 6, that pollen may pass along a food chain, being finally excreted by the carnivore at the end of the chain.

In the archaeological context the greatest attention has been devoted to coprolites of human origin. These are preserved in dry deposits, such as those in certain caves, and have been studied particularly in cave systems in the United States. Investigations have mainly embraced macroscopic remains of plant foods and have shed much light on the dietary habits of the occupants of the caves. In some cases pollen analyses have also been carried out, sometimes in the hope of obtaining palaeoenvironmental evidence, but more realistically to add to the palaeobotanical evidence of cultural practices. Martin and Sharrock (1964) have dramatically shown the possibilities of faecal analysis on samples from the Glen Canyon area of Utah. The pollen analyses gave very high percentages of species of cultural importance and relatively small amounts of regional and long-distance pollen. A very useful discussion of the potentialities of this line of investigation is given by Schoenwetter (1974), and another by Bryant (1974) in the same publication. A particularly interesting study was made of a prehistoric site on Wetherill Mesa, Colorado (Martin and Byers 1965). The pollen of some food plants was present in very low numbers in the occupational deposits (e.g., *Cucurbita*), but registered at higher levels in human coprolites from the site. Others were present at very high frequencies in the deposits (e.g., *Cleome,* bee-weed), suggesting that flower heads were discarded after food preparation. The authors quote other examples of pollen analyses of pueblo deposits, in which attempts have been made to obtain a picture of the background by excluding what are clearly culturally-concentrated taxa, in much the same way that aquatics or waterside species (e.g., *Alnus*) are often excluded from conventional pollen sequences. In general it seems that there is too much variability between samples to allow any reliable conclusions to be drawn about the environmental background, and therefore about dating. Much more could be learnt about the foods that the people ate; in some cases it seems that they were deliberately eating pollen-rich material such as young flowers, which would be rich in protein. If so, this of course gives some evidence of the season at which the cave was occupied. In fact good evidence of seasonality could sometimes be obtained, not only from the pollens that were present in such quantity that they must have been ingested deliberately, but also from pollen of apparently accidental origin, often of plants of no food interest. It was mentioned above that Bottema had found high percentages of the pollen of Liguliflorae in the nest of a bee, but Bryant (1974) also found high values for this taxon in some of the coprolites he analysed. Here there is no apparent connection with bees, and he concludes that flowers of dandelions were deliberately being eaten.

The possible effect of digger bees in concentrating pollen in soils has been discussed at some length by Havinga (1963:25). In sandy soils he found not only high concentrations ('arbitrary overrepresentation') of pollen types clearly out of context—e.g., *Valeriana* pollen in a heath soil—but he also found such concentrations associated with the burrows of digger bees (*Halictus*), and actually extracted a digger bee from a profile he was sampling. He points out that *Halictus* has a liking for the yellow-flowered composites, and so may be responsible for the 'arbitrary overrepresentation' of *Taraxacum*-type pollen, which can be such a mystery. However, he does describe such over representation as occasional, and only claims that digger bees may be a partial explanation.

From my own experience I would want to underline this caution. Havinga has been dealing mainly with sandy soils, and in all the years I have worked on sandy soils I have rarely, if ever, met comparable 'arbitrary overrepresentation'. Havinga mentions that evidence of digger bees has also been found in peats. On the other hand, the commonest context for overrepresentation of Liguliflorae pollen reflected in this book is the rendzina, either buried or modern, usually associated with chalk. Another important context is the cave deposit (see Ch. 6). It seems unlikely that digger bees, even if of several species, would be important influences in such a diverse range of habitats. I am left with a feeling—no more—that man himself is a more likely cause, but at the moment how he may have brought about this effect can only be a matter of speculation.

SETTLEMENT SITES

The foregoing discussion about human coprolites, which it seemed logical to include under the heading of pollen concentration by animals, indicates one way in which man himself can affect pollen spectra. There are, of course, many other ways, so that when man tends to concentrate these we may expect to find a great variability in the pollen spectra obtained. As suggested at the beginning of this chapter, the variability might preclude such evidence as a basis for palaeoenvironmental interpretation, but it could provide information of another sort, particularly cultural. For this reason, there may be positive value in investigating those sites where man has congregated, that is settlement sites. Nearly 30 years ago Troels-Smith (1955:11–58) and Welten (1955:61–88) drew attention to the fact that cultural layers of the pile-dwelling sites in the Swiss lakes showed pollen spectra differing in composition from samples obtained outside the cultural area. Welten (1967) presented a series of analyses from samples taken along a transect extending from a pile-dwelling site bounded by a palisade to a distance of 29 m to the landward from the palisade. His diagram showed a very marked contrast between samples from the dwelling place itself, which included good representation of a number of taxa which were scarcely represented outside the palisade, and the samples from beyond the palisade. In fact for many taxa it was remarkable how little influence the high values of the settlement site had on samples from a comparable level beyond the confines of the site. Both Troels-Smith and Welten had been looking for evidence of the impact of man on the vegetation immediately surrounding the sites investigated, and they did find clear evidence of this. However, it was apparent, particularly from Welten's later evidence, that an element of the settlement site pollen was not derived from the vegetation on or immediately adjacent to the site, but was derived from material brought in to the site, material that was deliberately selected and may have come from plant communities not represented along the lake margin.

For instance, leaves or forest litter had apparently been brought in, perhaps for bedding or floor covering, and with them came a suite of pollen types from woodland.

Somewhat similar conclusions were reached by Schweingruber (1976) with regard to another Swiss site, Egolzwil 5, a lakeside site in the Early Neolithic which is now covered with peat. Apart from the clear evidence of farming, there was some indication that certain plants had been brought onto the settlement itself by man. These included ramsons (*Allium ursinum*), flowering shoots of ash (*Fraxinus*), lime (*Tilia*), and maple (*Acer*), together with shoots of ivy (*Hedera*). No suggestion was made as to why these plants were introduced onto the site (the nature of these plants suggests that there was more than one reason), and it is interesting to note that they have different flowering seasons.

Pollen of *Allium*, an insect-pollinated herb, is not common in pollen analyses of any sort, but it was recorded in considerable quantities (32% of the total pollen) in the old land surface beneath a Bronze Age barrow on heathland in southern England (Scaife, in press). It is difficult to imagine the calcicolous ramsons as part of the ground flora on noncalcareous sand even in the Bronze Age, so the possibility that it was brought to the site by man should be considered in this context too.

Later settlement sites, particularly medieval ones, can provide abundant evidence of the activities that went on in or near the site. With the good preservation of organic matter that is often found on such sites, pollen is not usually the only or even the main source of information. Macroscopic plant remains and insect remains can sometimes give a remarkably detailed picture. Nevertheless, the presence of pollen, particularly where there seems to be overrepresentation of certain significant taxa, can fill in the general picture. Such a case was the early medieval site in York (Buckland *et al.* 1974). Here abundant remains of seeds and other plant remains occurred, together with a very rich invertebrate fauna. Pollen was also present, but the authors point out that the pollen spectrum was completely devoid of trees and shrubs and was dominated by grass and cereal pollen, contrasting with other contemporary spectra in which trees were significantly represented. They conclude, therefore, that this is not a spectrum derived from atmospheric pollen, but that the pollen had been brought into a building with plants used for floor covering, et cetera.

Recently, particular attention has been directed to urban sites ranging from Roman to medieval in age by the same team of co-workers (Greig 1982). Greig gives a detailed analysis of the ways in which pollen reaches and gets moved around in the urban environment. He strongly reinforces the point made at the beginning of this chapter that certain types of pollen assemblage are to be interpreted in terms of human activity rather than in terms of the background environment. He quotes some striking examples of overrepresentation in pollen spectra, the two main ones being (1) high values of Gramineae and Liguliflorae, which are seen as derived from natural transport and deposition (a view which I personally have some difficulty in accepting, as reference to the discussion on high Compositae values in the last chapter will explain); and (2) high values of Cerealia and occasionally other taxa, which clearly cannot be natural in origin and must be seen in terms of human agency. Greig supports his interpretation with some interesting observations on the association of pollen with plant foods. It had already been shown by Robinson and Hubbard (1977) that cereal pollen could be extracted from the bracts of hulled cereal grains, showing that pollen can be transported in unexpected ways and at times other than the flowering season, and they speculated that pollen may be brought onto a site in material for threshing. Greig takes this a stage further and analyses whole cereal plants from modern fields. They gave surprisingly high counts of cereal pollen grains from the spikes of wheat and barley, ranging from 30–40,000 grains per ear, whilst it was estimated that a panicle of oats could carry some 45,000 grains. The straws also yielded pollen, predominantly of grasses and cereals, but with other associated taxa such as *Urtica* (nettle) and *Papaver* (poppy). Other food plants showed much less association of pollens with the fruits. However, even processed foods were found to contain pollen. Dickson (1978) had previously shown that honey can yield pollen, and has suggested that organic remains found in a vessel were of honey, for they contained pollen of *Tilia* (lime).

Greig found cereal pollen to be present in considerable quantities in crushed crystal malt and in oatcakes, and traces could be found in wholemeal flour and rye bread. Most surprising was the identification of pollen of cereal and *Vitis* (grape) in a currant tea-cake (the so-called currants are not true currants but come from a small black grape). It has been shown that the pollen in foods such as barley meal survives passage through the human digestive tract.

As we have seen, Greig and his co-workers concluded that the pollen spectrum at their York site comes from material not exposed to the open air, because of the absence of tree pollen. Had there been such exposure one would expect some apparent influence of the pollen rain from outside. However, even a closed building is rarely pollen-proof; contamination of samples within a building with contemporary pollen is a hazard all pollen analysts are aware of. I myself had the cautionary experience of finding that the pollen suite obtained from a mesolithic barbed point was dominated by the pollen of London plane (*Platanus* ×*hispanica*). This factor makes it all the more remarkable that the York material should have been so little influenced in this way, but it could well be a matter of statistics. If one is examining a sample poor in pollen the outside influence will be concentrated, but if the sample has a rich pollen content, a low level of contamination may not show up significantly in the count.

In this discussion of the pollen analysis of urban sites no account has been taken of the fact that other biological evidence may be present and may in fact greatly outweigh that of the pollen itself. Macroscopic remains of plants—leaves, seeds, fruits, etcetera—and particularly insect remains have been used as a basis for interpretation, and in some cases such data are critical in the interpretation of the pollen analyses. An example of such a multidisciplinary approach is the investigation of the so-called dark earth which is a prominent but unexplained feature of urban sites from Roman to medieval times. Its origin has been the subject of much archaeological speculation (some of it very ill-informed), but an approach combining soil science and botanical analysis (including pollen analysis) is at last giving some hard data on which archaeologists can build (MacPhail 1981). Our present purpose, however, is merely to demonstrate the potential for pollen investigation in such contexts.

Another example of a medieval site analysed for pollen was a barrel latrine at Worcester, also investigated by Greig (1981). As usual in such rich sites, pollen analysis only plays a secondary part in the investigation, but this piece of work is noteworthy because Greig discusses at some length the possible sources of organic waste in a medieval homestead. As at York, the pollen spectrum was dominated by pollen of wild grasses and cereals in almost equal proportions.

An unusual pollen assemblage was extracted from the organic fill of a drain in the grounds of Mount Grace Priory, Yorkshire, a late fourteenth-century Cistercian foundation. From a list of 15 taxa and a total count of 295 grains, 80% proved to be of yew (*Taxus*), a wind-pollinated species that one might expect to be abundant in the grounds of a religious house. The concentration in a drain, however, could be due as much to storm water as to wind dispersal (Dimbleby 1970).

It is apparent from these examples that pollen spectra from settlement sites of any age can differ markedly from contemporary spectra obtained from sites not associated with human activity; in other words, the pollen spectra are not in equilibrium with the pollen rain of the time. It follows, therefore, that such spectra are of minimal use for ecological interpretation. Nevertheless, they may tell us a good deal about the activities of man on settlement sites, particularly if combined with other forms of botanical and zoological evidence. There is clearly a need for more work on the means by which pollen is transported through human agencies and the ways in which it can become preserved and even concentrated in organic deposits on occupation sites.

EVIDENCE OF HUSBANDRY

The material I want to discuss under this heading cannot be clearly demarcated from what follows under the heading of agriculture, but I have to separate it because it can be demonstrated even on pre-neolithic sites. The term husbandry here is used to describe a relationship between man, plants, and animals which involves manipulation and exploitation of the environment but which falls short of agriculture involving recognizable domesticates.

Repeated reference has been made to the high concentrations of ivy (*Hedera*) pollen found in some open sites of mesolithic age, and it was argued that this phenomenon was related in some way to husbandry of the red deer. It is significant that most examples of pollen spectra showing high *Hedera* frequencies are occupation sites of one sort or another (Simmons and Dimbleby 1974). Spectra from non-occupation sites reflect the low values one would expect for a taxon in which entomophily is the means of cross-pollination. It appears that the ivy pollen is concentrated on occupation sites in some way, perhaps by the collection of flowering ivy for winter feed, or perhaps even by the collection of the droppings of deer which have been feeding on ivy. Such concentrations are not, however, confined to ivy. At the mesolithic site of Montbani, Planchais (1972; Parent and Planchais 1972) has found high frequencies of the pollen not only of ivy but also of Asteraceae (Tubuliflorae), Papilionaceae (cf. *Genista*) and Umbelliferae (cf. *Daucus*), all entomophilous types (see Fig. 34). Another umbelliferous species, *Sanicula europea*, like *Hedera*, is a woodland plant, and is moreover a calcicole. It could not, therefore, have come from the immediate locality, which was one of acid sand, suggesting that a more distant forest environment was being exploited, perhaps as a source of fodder. Much depends here on the precision of the pollen identification, but it is apparent that pollen analysis in such contexts has the potential of providing information about cultural activities which could not be extracted from the purely artifactual data.

AGRICULTURE

There is, of course, no reason why overrepresentation of wild species which are potential fodder plants should not occur in later periods, and indeed this seems to be the case. I have come across high frequencies of *Hedera* pollen in Bronze Age contexts; one such has already been mentioned above, in the round barrow at Portesham (see Ch. 4).

Wild vegetation was not only exploited for fodder, but served a number of other agricultural purposes. One obvious example is the use of cut plant material as floor covering, thatch, or litter in animal stalls; such plant material is in due course disposed of and replaced. As Greig has suggested in his account of the barrel latrine at Worcester, disposal of organic rubbish may be *via* the latrine, and of course it may not end there either. Pollen grain walls are impregnated with one of the most resistant plant materials, sporopollenin, sometimes allowing the recognition of pollen grains when the other recognizable materials no longer persist. In a farmstead it is to be expected that organic waste, including human excreta, eventually finds its way onto the fields as manure. This then points the contrast between an occupation or settlement site in the restricted sense, and the associated field system under the control of that settlement. It is only relatively recently that archaeologists have come to realize that a rural settlement without a field system is incomplete, though there has been more awareness that the reverse is also true. How many excavations of Roman villas have extended beyond the actual building complex of the villa? Recent work by Goodburn at Winterton, Humberside, is showing that archaeological evidence of the associated fields can be found if looked for, and indeed can repay detailed study.

However, this does raise anew the question of what is a site, and it must be obvious that a great deal of the landscape has been intensively used by man even though there is no artifactual evidence to prove it. Sometimes evidence of use can be thrown up by pollen analysis, as for instance in the pollen analysis of chalk soils in the south of England, a centre of intensive prehistoric agriculture. As explained in Chapter 1, calcareous soils offer a poor prospect for pollen analysis, but in some cases adequate pollen counts can be obtained, albeit at low absolute frequencies, and, as already pointed out, soil fauna may destroy any pollen stratification in the soil profile. Pollen spectra from such soils commonly show the same group of taxa predominating, namely Gramineae, *Plantago lanceolata,* Liguliflorae, *Corylus* (hazel), and *Pteridium* (bracken). The surprising member of this group is the last, for it does not occur as a normal component of the chalk vegetation today. Whilst it is not a complete calciphobe, for it can be found on the Cotswold limestones, it is more characteristic of neutral to acid soils, particularly if they have had a woodland history. Wherever it is found in the chalk countryside it is on decalcified deposits such as the Clay-with-Flints. Why, then, should its spores be such a characteristic feature of soils developed on chalk? Airborne transport of the spores seems to be ruled out, partly on the grounds that other airborne microspores are not found in the same way (though admittedly bracken spores are very resistant to decay), but largely because such contamination from outside is almost invariably swamped by the pollen component from the vegetation growing on or adjacent to the site samples. A glance at the two diagrams, one a buried soil beneath the South Street long barrow, Fig. 35), and the other from an open soil profile on Overton Down (Fig. 36; Jewell 1963) shows the high proportions that can be reached by the bracken microspores.

One possible explanation of this phenomenon is that the spores have been introduced by human activity, probably through manuring of the land with

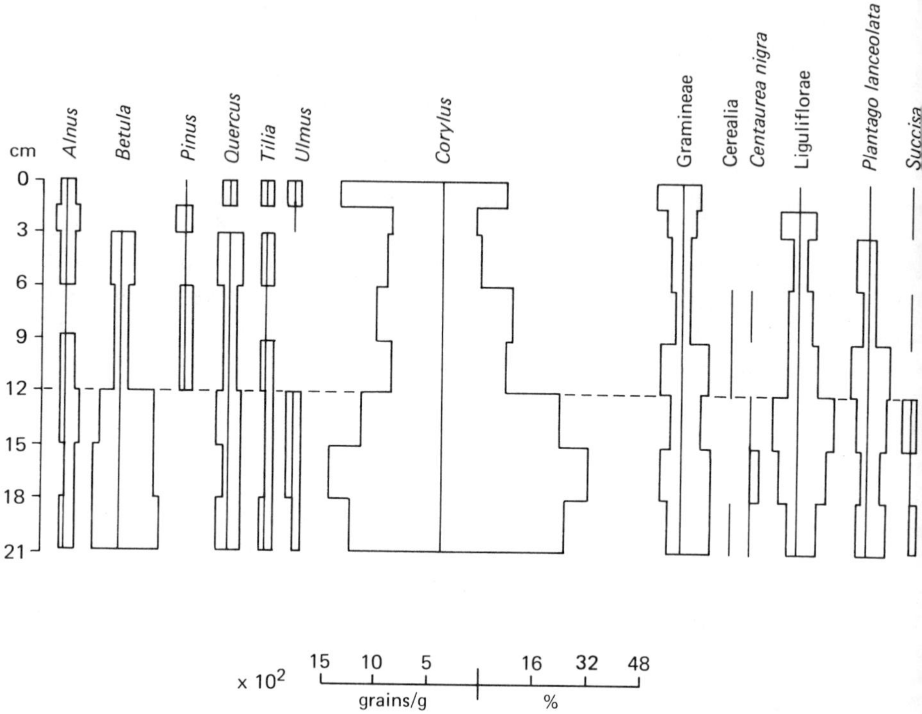

Figure 35. Pollen profile through an old land surface, South Street long barrow, Wiltshire.

farmyard manure and household refuse. The presence of small-sized sherds and fragments of charcoal in such contexts seems to confirm that such manuring took place in prehistoric times, and microscope preparations made for pollen analysis sometimes contain charred tracheids of ferns, some of which seem to correspond closely to bracken (Dimbleby and Evans 1974), suggesting that bracken fronds had been in contact with fire, as had the charcoal fragments mentioned above. Bracken is a plant widely used for animal bedding; it is poisonous to cattle in the green stage, but once it turns colour it loses its toxicity and of course this is the time at which the sori mature and the spores are released. Bracken can also be used for thatching and in hilly chalk country, well above reedswamp levels, there are few other plants suitable for this purpose, so it might be collected from areas of acid drift or Clay-with-Flints.

Whilst this explanation accounts for the presence of bracken spores in the calcareous soils, it is not possible to apply the same reasoning to the other dominant taxa, for they could all have originated from vegetation growing on chalk soils. The high proportion of Liguliflorae pollen is very characteristic of

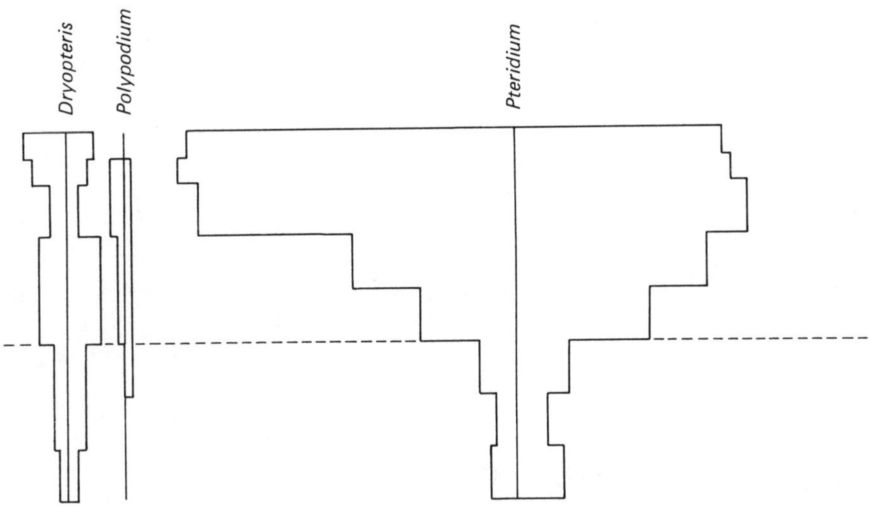

Figure 35. *continued*

base-rich soils. There may be some overrepresentation in the pollen counts; this pollen type seems to be resistant to decay, and is also readily recognizable even when badly corroded or damaged. Nevertheless, as pointed out earlier, differential preservation cannot increase numbers, and this taxon is insect-pollinated and therefore not a massive contributor to the pollen rain. Concentration by ground-nesting bees was referred to earlier in this chapter, but whilst this could explain the occasional high count it is difficult to see how it could be the cause of such a regular and widespread phenomenon.

It is to be expected that samples from sites in an agricultural environment will have above average percentages of pollen of weeds and ruderals. Some of these, like the Liguliflorae just referred to, are entomophilous, and other such as *Plantago* spp. are anemophilous, but their relative abundance does not always reflect the differences in pollen production and dispersion. Turner (1964) devised the so-called arable-pasture index based on the occurrence of cereal pollen and the relative proportions of weeds which are characteristic of arable and pasture conditions respectively. The ratio is given as: *Plantago* ÷ (*Plantago* + Compositae + Cereal + Cruciferae + *Artemisia* + Chenopodiaceae). Turner suggests that a value of this ratio below 15% indicates arable and over 50% is pasture. Unfortunately pollen identification often can not be taken below the level of family, and that level may include species of different ecological tolerances. For example the Ranunculaceae, regarded by some as

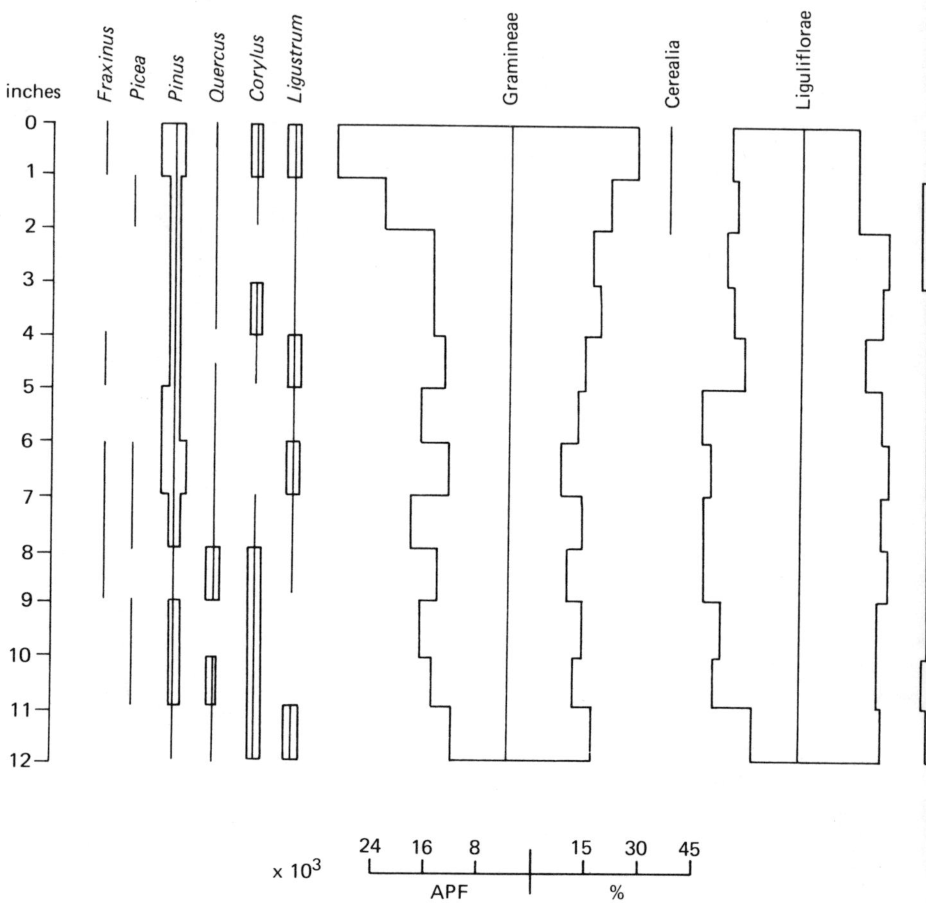

Figure 36. Soil pollen diagram from Overton Down, Wiltshire (1958).

an 'arable' taxon, includes *Ranunculus repens,* a plant of disturbed soil, and *R. bulbosus,* characteristic of pasture. Another drawback is that a species may be 'characteristic' of a habitat yet be able to occur in very different habitats. I have seen *Plantago lanceolata,* a plant 'characteristic' of pasture, growing vigorously in the disturbed sand of a sand-pit. Applying this ratio to the ploughsoils beneath the ship barrow and Mound 5 at Sutton Hoo (Bruce-Mitford 1975), a value of 42% is obtained for the former and 52% for the latter. Both of these are well above Turner's value of 15% for arable, yet these are both buried plough soils and contain some of the highest frequencies of cereal pollen that I have come across in soils. It seems that such a ratio is not

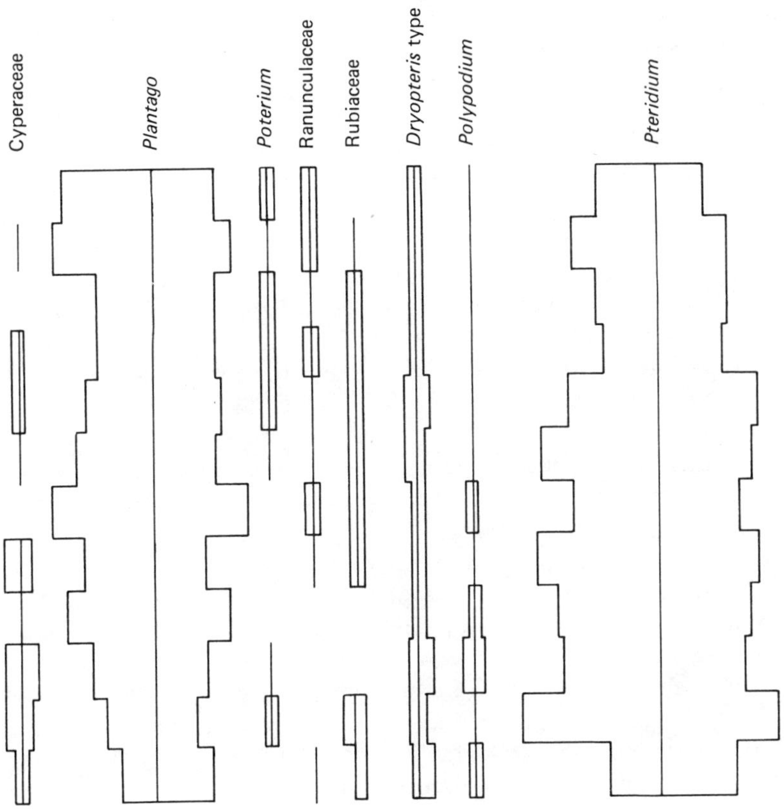

Figure 36. *continued*

applicable to pollen analyses of soil horizons, where the bulk of the pollen has come from plants growing on or adjacent to the sampling point, in contrast to samples from bogs where the pollen has reached its destination by transport often over a considerable distance.

The high values for cereal pollen at Sutton Hoo (Fig. 15) again raise the question whether this pollen all derives from the plants growing on the site, or whether there are often other sources which could contribute to the soil pollen. Unfortunately, no attempt was made to identify the cereal pollen, which it is now seen could have been an important point. If the pollen is all derived from the growing plants, it might be argued that the cereal was rye (*Secale*) which is the only fully anemophilous species. All the others release their pollen more-or-less cleistogamously, only a few anthers on each ear releasing pollen into the air. However, as we have seen it has been shown that pollen grains may remain attached to hulled cereal grains, and to other parts of the spikelet too. More recently Lambrick and Robinson (1979) have described a waterlogged sample which gave a cereal pollen percentage of 60% of total pollen. This sample was a laminated peat composed almost entirely of wheat chaff, the bulk of it from *Triticum spelta*. This is very significant for it shows that cereal pollen can remain behind after the ears have been threshed and the grain extracted. If such refuse is then incorporated into farmyard manure it is obvious that it could find its way back onto the fields and so add to the pollen content of the soil. It is a matter of conjecture what other pollens may follow the same route. A further complication is that farmyard manure is not only applied to arable soils. It is sometimes applied as a top dressing on pasture. If this is an old practice the pasture soil may well be receiving a dose of cereal pollen through this process. Animals that are partially stall-fed and are then turned out onto pasture could bring about a similar effect through their dung.

A rather parallel practice is found today in Switzerland. The *Fettwiesen*—valley meadows with an extraordinarily rich flora—are produced by continual dunging, until an almost peat-like organic soil is built up. In a montane country with a climate that requires the winter stalling of animals, it is understandable that the accumulated manure will be put to good use, and with a limited arable acreage because of the terrain, meadows and pasture will be the obvious recipients.

This may give a clue to the use of manure in prehistoric times, particularly in an economy with a large pastoral element and a limited arable one. Farmyard manure is not only a valuable product, but if not regularly disposed of too much of it may become a nuisance in itself. It would be a natural answer to get rid of this waste product wherever possible, and if it added to fertility this would be a bonus. The very high percentages of pollen and spores in chalk and other pasture soils could be the last vestige of a practice which had a dual purpose—to get rid of the manure, and, perhaps secondarily, to boost fertility.

A similar explanation may account for the anomalously high count of Cerealia pollen found in a sample from the modern land surface at Ranscombe, on the South Downs (see Ch. 4, Table 2). This surface was not under cultivation at the time. In the light of the above discussion animal dung—horse droppings or stable manure perhaps being the most likely—could have introduced a biased representation of this particular pollen type.

I am also reminded of a pollen analysis published by Rodden (1962) from the early neolithic site of Nea Nikomedeia, Greece, which also produced very high percentages of cereal pollen. There was no cultural reason why cereals should not be an important element in the economy of this time, of course, but some pollen analysts regarded such high values as ‘unacceptable’. It is usually pointed out that there are some wild grasses which have grains very similar in size to those of cereals, so that without critical study there was the possibility of confusion (Leroi-Gourhan 1969). Indeed, this is one reason why the term Cerealia came into regular use, for under its umbrella grains of 'cereal-type' could be recorded without the obligation to distinguish them further. However, in view of the more recent evidence presented here there is no longer any need to 'explain away' all high Cerealia counts; they may indeed be exactly what they seem, but of course they are not in equilibrium with the contemporary pollen rain.

FLORAL TRIBUTES

Where floral material is collected together it is obvious that the pollen will also be introduced in large quantities. I have never investigated the soil pollen in a graveyard, but it could produce surprising results. Interestingly there is one such record from archaeology, namely the Neanderthal burial in the Shanidar Cave (Leroi-Gourhan 1975). Associated with the skeleton were the remains of several species of brightly coloured flowers, apparently worked into *Ephedra* branches. Samples of sediment associated with the burial were analysed for pollen, revealing the species whose flowers were identifiable and others, mostly medicinal herbs, present in the neighbourhood even now. Many of the pollen grains occurred in large clusters; they had clearly been deposited in the soil as the anthers decayed and had never been dispersed.

Clusters of pollen are not uncommonly found in soil pollen preparations and should always be recorded. It is apparent that a large cluster is not likely to be carried far by the wind and is most likely to have come from a fallen flower or anther, in which case it is a strong indication that the plant was growing near (or had been brought to) the site from which the sample came.

In Chapter 4 reference was made to the distribution of Liguliflorae pollen at the constructional interface C/D in the little late Bronze Age barrow at Burley in the New Forest (Fig. 18). In lighter moments it has been suggested that this represents a floral tribute of some sort laid at this construction phase of the barrow. This is the nearest I have come to finding pollen evidence of the last rites and is an appropriate point at which to end this chapter.

POSTSCRIPT

The overall impression gained from this book may be that site pollen analysis is too complex to be a useful tool in archaeology. Such a judgment may stem from a belief that to be really useful the results of the technique should be simple and straightforward. In view of the complexity of human activities, and not infrequently their apparent irrationality, any straightforward outcome would have to be regarded with suspicion; to imagine otherwise would be naive. On the other hand, I take the view that the considerable variability is a positive factor which, given further pursuance, could add to our knowledge about the activity of prehistoric (or even later) man.

In the course of this book I have had cause to remark at some points that though pollen may be present and extractable from some deposits, it may not be possible to interpret it because of the doubts about the provenance of the pollen, in which circumstances pollen analysis is not worth applying. This is not the same as saying that where interpretation is difficult despite the apparently clear provenance of the pollen in the material, analysis is pointless. In my own experience I have seen a number of cases where analyses have proved difficult to interpret at the time, but in due course new light has been shed which has suggested a previously unconsidered line of explanation; because they are unexpected, such explanations are often the most valuable in putting a new angle on old questions.

The repeated occurrence of certain types of spectra in comparable contexts cannot be random and accidental; there must be some important factors governing the deposition of pollen, for instance in cave deposits, though it must be admitted that little critical research has been devoted to finding out what these might be. It is true of some of the ground covered by this book that the main reason for carrying out pollen analysis on the deposits is that pollen was present in them. By the nature of things it is often impossible to find modern conditions which parallel those likely to have pertained in the past and which could be investigated experimentally to provide the necessary basis for interpretation. Where they occur, opportunities should not be missed, such as with the reconstructions of prehistoric settlements on which appropriate activities have been carried out. The Iron Age farmstead at Butser Hill (Reynolds 1976) or the site of the BBC TV programme *Living in the Past* (Percival 1980), come to mind as two possibilities.

It could be argued that since there are so few possible checks on the bases of interpretation there is no point in carrying out such an uncertain application of a well-founded technique. But it must be acknowledged that conventional pollen analysis itself developed in much the same way. It too was based in the first place on the fact that pollen was present in the stratified deposits, and interpretations were made with little understanding of how the pollen got into the deposits. Detailed pollen analyses of lake deposits were being made even before the Second World War, yet it was not until the 1960s that attention was seriously directed to air movements over lakes or bogs, and it was a decade later that it was shown that the major input of pollen into a lake seems to be *via* streamflow and only a minor component comes from direct pollen rain onto the lake surface itself (Peck 1973; Bonny 1978).

Even more recently the very basis on which pollen diagrams are interpreted has come under scrutiny (Edwards 1979). This concerns in particular the basis of interpretation of clearance and regeneration phases as reflected in diagrams from stratified deposits; the demarcation between the boundaries of such events often seems to have been determined subjectively and has often been influenced by the concept of 'landnam' as originally put forward by Iversen (1941). Edwards seems to doubt whether landnam in these terms is commonly, if ever, found. Another factor which occurs to me, and which is seldom discussed, is the extent to which so-called pastoral phases were perpetuated by wild herbivores, rather than domestic ones. The question is probably unanswerable, which may be why it is never asked. Once a rich ground vegetation is released by the removal of overhead canopy, it may be exploited by domestic stock, but equally it will attract wild herbivores which will remain as long as the area provides suitable and accessible grazing. Buckland and Evans (1984) have discussed the pollen evidence for the perpetuation of clearings, made by grazing pressure by wild and domestic herbivores. Today we do not have wild herds on the same scale as in prehistoric times, though herds of feral deer can sometimes exert this impact even in an intensively farmed landscape; and we need not ignore the humble rabbit.

Having suggested that even conventional pollen analysis has its Achilles heel, it would be wrong to detract from the very positive results which have accrued from it. In the Preface to this book I remarked on the mass of evidence—indirect though it may be—that has come from such work regarding the activities of prehistoric man. Pollen data are the major source of our knowledge about the changes in the landscape brought about by the hand of prehistoric man. No doubt adjustments will have to be made in detail, but the broad groundwork seems incontrovertible. Nevertheless, the stratified deposits do only offer indirect evidence of human activities, so there is good reason for persevering with the palynology of archaeological sites in the hope of obtaining more direct correlations. Any information that emerges will be complementary to that obtained from stratified deposits, and it could well provide some detail of practices which could hardly be guessed at from the indirect evidence. The somewhat confusing body of data which has emerged so far will be honed down in due course until recognizable positive features appear.

One further point seems to me to be important in interpretation. Edwards (1979) sees the pollen analyst as underestimating the variability and complexity of prehistoric man's activities in the landscape, and the archaeologist sometimes attempts interpretations on inadequate ecological knowledge. Too often species are regarded as indicators, without realizing the range of tolerance many species have. The presence of *Plantago lanceolata* pollen may indicate farming, particularly pastoral farming. On the other hand, it may not. Many pollen analysts equate significant counts of ivy (*Hedera*) pollen with woodland, but a walk in the countryside soon casts doubt on this assumption. I know of at least one case where high counts of *Pinus* pollen have been interpreted as Boreal in age, when in fact they reflected the spread of pine in the last few centuries.

Such rules of thumb are dangerous, and nothing replaces a sound field knowledge of the ecology of the taxa concerned. Above all it is necessary to think not in terms of indicator species (insofar as they can be identified from their pollen alone), but in terms of communities and ecosystems. This is particularly important with mixed pollen assemblages which may derive from more than one ecosystem, a very common situation in the archaeological context. This is also the point at which links should be made with evidence from other disciplines where conditions allow, thus giving a potentially more complete picture of the ecosystem.

Appendix: SOIL POLLEN ANALYSIS METHOD

1. Collect about a matchboxful of soil for each sample, 5″ × 3″ polythene bags being very convenient for this purpose.
2. Dry as soon as possible, preferably in an oven at 90–100°C.
3. Crush lightly with a pestle and mortar, then sift through a 1 mm sieve.
4. Weigh out from 0.5 gm (surface and organic layers) to 4 gm (mineral subsoil) of soil into a 100 ml conical flask.
5. Add approximately 20 ml 10% sodium hydroxide and simmer gently on a hot plate for 40–60 minutes, the longer time being for clays and organic soils. Agitate occasionally to prevent soil from sticking to the flask.
6. Decant through a gauze filter (72 mesh) into 50 ml Polypropylene centrifuge tubes; centrifuge, decant, wash, centrifuge, and decant.
7. Add about 10 ml of 40–60% hydrofluoric acid and carefully bring to the boil in a water bath for 20–30 minutes, stirring frequently with a Polypropylene or P.T.F.E. stirring rod.
8. Transfer to 15 ml Polypropylene centrifuge tubes, centrifuge, and decant.
9. Add 10 ml 10% hydrochloric acid and warm in water bath for a few minutes, but do not boil. Centrifuge and decant.
10. Dehydrate with about 5 ml glacial acetic acid. Centrifuge and decant.
11. Add about 5 ml freshly made up acetolysis mixture (to 9 parts acetic anhydride, add 1 part conc. sulphuric acid drop by drop from a burette, vigorously agitating the mixture all the time). Bring to the boil, continue boiling until the mixture turns brown. Centrifuge and decant.
12. Add a few ml of glacial acetic acid; centrifuge, decant, wash, centrifuge, and decant again.
13. Add about 5 ml 10% sodium hydroxide and boil gently in a water bath for 10–15 minutes, centrifuge, decant, and wash twice.
14. Make up to 2 or 5 ml (the smaller volume if the material is poor in pollen) with 1:1 glycerol and water to which 0.5% safranin has been added in the proportions 1:50. If a lot of carbon or organic matter is present it may be necessary to add more safranin stain.
15. Using a calibrated dropper (e.g., hypodermic syringe), place one drop on a microscope slide and cover with a cover-glass.

REFERENCES

Armitage, P. L.
1975 The extraction and identification of opal phytoliths from the teeth of ungulates. *Journal of Archaeological Science* 2:187–197.

Arthur, J. R. B.
1965 Special note on the plant remains from Tonge, Kent. In Tonge medieval manor: Summary of results, edited by D. M. Ford. *Archaeologia Cantiana* 80:268.

Ashbee, P., and G. W. Dimbleby
1959 Excavation of a round barrow on Chick's Hill, East Stoke Parish, Dorset. *Proceedings of the Dorset Natural History and Archaeological Society* 80:146–159.
1976 The Moor Green barrow, West End, Hampshire. *Proceedings of the Hampshire Field Club and Archaeological Society* 31:5–18.

Ayers, B., and P. Murphy
1983 A waterfront excavation at Whitefriars Street car park, Norwich, 1979. *East Anglian Archaeology Report* 17.

Balaam, N. D., K. Smith, and G. J. Wainright
1982 The Shaugh Moor Project: Fourth Report—Environment, Context and Conclusion. *Proceedings of the Prehistoric Society* 48:203–278.

Ball, D. F.
1975 Processes of soil degradation. In The effect of man on the landscape: the Highland Zone, edited by J. G. Evans, S. Limbrey, and H. Cleere, pp. 20–27. *Council for British Archaeology Research Report* 11.

Bandi, H. G., C. Ludin, W. Manber, S. Schaub, E. Schmid, and M. Welten
1954 Die Brügglihöhle an der Kohlholzhalde bei Nenzlinger (Kt. Bern), eine neue Fundstelle des Spätmagdalenien im unteren Birstal. *Jahrbuch des Bernischen Museums in Bern* 1953/1954:47–76.

Bastin, B.
1979 L'analyse pollinique des stalagmites: une nouvelle possibilité d'approche des fluctuations climatiques du Quaternaire. *Annales de la société géologique de Belgique* (1978) 101:13–19.

Beaulieu, J. L. de
1967 Aspects de la végétation d'un nouveau site préhistorique mindelien des Alpes maritimes, d'après l'analyse pollinique. *Paris Comptes rendus de l'Académie des Sciences* 264, Series D:2741–2743.

Behre, K-E.
1969 Untersuchungen des botanischen Materials der frühmittel-alterlichen Siedlung Haithabu. In *Bericht über die Ausgrabungen in Haithabu: Bericht 2*, edited by K. Schietzel, pp. 7–55. Wachholtz, Neumünster.
1976 *Die Pflanzenreste aus der frühgeschichtlichen Wurt Elisenhof*. Lang, Bern.

Bell, M.
1982 The effects of land-use and climate on valley sedimentation. In *Climatic change in later prehistory*, edited by A. F. Harding, pp. 127–142. The University Press, Edinburgh.
Biek, L.
1963 *Archaeology and the microscope*. Lutterworth Press, London.
Birks, H. J. B., and H. H. Birks
1980 *Quaternary palynology*. Arnold, London.
Birley, R.
1977 *Vindolanda*. Thames and Hudson, London.
Bonny, A. P.
1978 The effect of pollen recruitment processes on pollen distribution over the sediment surface of a small lake in Cumbria. *Journal of Ecology* 66:385–416.
Boon, G. C.
1980 Caerleon and the Gwent Levels in early historic time. In *Archaeology and coastal change*, edited by F. H. Thompson, pp. 24–34. Society of Antiquaries, London.
Bottema, S.
1966 Palynological investigation of a settlement near Kalopsidha (Cyprus). In Excavations at Kalopsidha and Ayios Iakovos in Cyprus, edited by P. Åström *et al.*, *Studies in Mediterranean Archaeology* II, pp. 133–134.
1975 The interpretation of pollen spectra from prehistoric settlements (with special attention to Liguliflorae). *Palaeohistoria* 17:17–35.
Bowen, H. C., and P. J. Fowler
1963 Earthwork circles and mounds on Studland Heath, Dorset. *Antiquity* 37:220–223.
Brisbane, M., and S. Clews
1979 The East Moor Field Systems, Altarnun and North Hill, Bodmin Moor. *Cornish Archaeology* 18:33–56.
Bruce-Mitford, R.
1975 *The Sutton Hoo Ship-Burial* (Vol. 1). British Museum, London.
Bryant, V. M.
1974 Pollen analysis of prehistoric human feces from Mammoth Cave. In *Archaeology of Mammoth Cave Areas*, edited by P. J. Watson, pp. 203–209. Academic Press, New York.
Buckland, P. C.
1976 The environmental evidence from the Church Street Roman Sewer System. In *The Archaeology of York* 14/1, pp. 1–44. Council for British Archaeology, London.
Buckland, P. C., and K. J. Edwards
1984 The longevity of pastoral episodes of clearance activity in pollen diagrams: the role of post-occupation grazing. *Journal of Biogeography* 11:243–249.
Buckland, P. C., J. R. A. Greig, and H. K. Kenward
1974 York: an early medieval site. *Antiquity* 48:25–33.
Burrin, P. J. and R. G. Scaife
1984 Aspects of Holocene valley sedimentation and floodplain development in southern England. *Proceedings of the Geologists' Association* 95:81–96.
Butzer, K. W.
1971 *Environment and archaeology*. 2nd Edition. Aldine-Atherton, Chicago.
Case, H. J., G. W. Dimbleby, G. F. Mitchell, M. E. S. Morrison, and V. B. Proudfoot
1969 Land use in Goodland Townland, Co. Antrim, from Neolithic times until today. *Journal of the Royal Society of Antiquaries of Ireland* 99:39–53.
Caseldine, A. E.
1983a Problems of relating site and environment in pollen-analytical research in south-west England. In Site, Environment and Economy, edited by B. Proudfoot. *British Archaeological*

Reports (International Series) 173:61–74.
1983b Recent palaeobotanical investigations at Meare Village West, Somerset. In Site, Environment and Economy, edited by B. Proudfoot. *British Archaeological Reports (International Series)* 173:75–84.

Casparie, W. A. and W. Groenman-van Waateringe
1980 Palynological analysis of Dutch barrows. *Palaeohistoria* 22:7–65.

Clark, J. G. D.
1954 *Excavations at Star Carr, an early mesolithic site at Seamer, near Scarborough, Yorkshire.* The University Press, Cambridge.

Cotton, M. A., and S. S. Frere
1968 Ivinghoe Beacon excavations 1963–1965. *Records of Buckinghamshire* 18:250–251.

Damblon, F.
1974 Nouvelles recherches à la grotte de Remouchamps. *Bulletin de la société royale Belge d'anthropologie et de préhistoire* 85:132–155.

Darwin, C.
1881 *The formation of vegetable mould through the action of worms.* Reprint edition, Faber and Faber, London, 1956.

Dauber, A., A. Fietz, and G. Lang
1955 Römische Brunnen in Pforzheim. *Beiträge zur naturkundlichen Forschung im Südwestdeutschland* 14:43–56.

Dickson, J. H.
1978 Bronze Age mead. *Antiquity* 205:108–113.

Dickson, J. H., C. Dickson, and D. J. Breeze
1979 Flour or bread in a Roman military ditch at Bearsden, Scotland. *Antiquity* 53: 47–51.

Dimbleby, G. W.
1950 *The ecology of some British podzol formations.* Unpublished D. Phil. dissertation, Oxford University.
1952 Soil regeneration on the North-east Yorkshire Moors. *Journal of Ecology* 40:331–341.
1955a Ecological study of buried soils. *Advancement of Science* 12:11–16.
1955b Pollen analysis as an aid to the dating of prehistoric monuments. *Proceedings of the Prehistoric Society* 20:231–236.
1957 Pollen analysis of terrestial soils. *New Phytologist* 56:12–28.
1961a Soil pollen analysis. *Journal of Soil Science* 12:1–11.
1961b Transported material in the soil profile. *Journal of Soil Science* 12:12–22.
1962a The development of British heathlands and their soils. *Oxford Forestry Memoirs* 23.
1962b Soil pollen distribution in relation to organic horizons. *Welsh Soils Discussion Group* Report No. 3.
1963a Pollen analyses from two Cornish barrows. *Journal of the Royal Institution of Cornwall* (n.s.) 4:364–375.
1963b Pollen analysis of a mesolithic site at Addington, Kent. *Grana Palynologica* 4: 140–148.
1969 Pollen analysis. In *Science in Archaeology*, 2nd ed., edited by D. Brothwell and E. Higgs, pp. 167–177. Thames and Hudson, London.
1970 Unpublished analytical data from Mount Grace Priory. Filed at the Institute of Archaeology, London University.
1976–1977 A buried soil at Innisidgen, St. Mary's, Isles of Scilly. *Cornish Studies* 4/5:5–10.
1978a *Plants and Archaeology*, 2nd Edition. Baker, London.
1978b A pollen analysis from Verulamium (St. Michael's 1966). *Hertfordshire Archaeology* 6:112–115.

1978c Changes in ecosystems through forest clearance. In *Conservation and Agriculture*, edited by J. G. Hawkes, pp. 3–16. Duckworth, London.
Dimbleby, G. W., and R. J. Bradley
1975 Evidence of pedogenesis from a Neolithic site at Rackham, Sussex. *Journal of Archaeological Science* 2:179–186.
Dimbleby, G. W., and J. G. Evans
1974 Pollen and land-snail analysis of calcareous soils. *Journal of Archaeological Science* 1:117–133.
Dimbleby, G. W., and M. C. D. Speight
1969 Buried soils. *Advancement of Science* 26:203–205.
Dimbleby, G. W., J. R. A. Greig, and R. G. Scaife
1981 Vegetational history of the Isles of Scilly. In Environmental aspects of coasts and islands, edited by D. Brothwell and G. W. Dimbleby. *British Archaeological Reports (International Series)* 94:127–144.
Drewett, P. L.
1975 The excavation of a turf barrow at Minsted, W. Sussex, 1973. *Sussex Archaeological Collections* 113:54–64.
Edwards, K. J.
1979 Palynological and temporal inference in the context of prehistory, with special reference to the evidence from lake and peat deposits. *Journal of Archaeological Science* 6:255–270.
Edwards, K. J., and K. R. Hirons
1984 Cereal pollen grains in pre-elm decline deposits; implications for the earliest agriculture in Britain and Ireland. *Journal of Archaeological Science* 11:71–80.
Evans, J. G., and S. Limbrey
1974 The experimental earthwork on Morden Bog, Wareham, Dorset, England: 1963–1972. *Proceedings of the Prehistoric Society* 40:170–202.
Evans, J. G., and K. W. G. Valentine
1974 Ecological changes induced by prehistoric man at Pitstone, Buckinghamshire. *Journal of Archaeological Science* 1:343–351.
Faegri, K.
1962 Palynology of a bumble-bee nest. *Veröffentlichungen des Geobotanischen Institutes Rübel in Zürich* 37:60–67.
Faegri, K., and J. Iversen
1975 *Textbook of pollen analysis* (3rd ed.). Blackwell, Oxford.
Fowler, P. J.
1983 *The farming of prehistoric Britain*. The University Press, Cambridge.
Frenzel, B.
1964 Zur Pollenanalyse von Lössen. *Eiszeitalter und Gegenwart* 15:5–39.
Girard, M., and J. Renault-Miskovsky
1969 Nouvelles techniques de préparation en palynologie appliquées a trois sédiments du quaternaire final de l'Abri Cornille (Istres, Bouche du Rhône). *Bulletin de l'association française pur l'étude du Quaternaire* 1969:275–284.
1979 Analyse pollinique de la grotte de Coupe-Gorge à Montmaurin (Haute-Garonne). *Bulletin de l'association française pour l'étude du Quaternaire* 1979:175–189.
Godwin, H.
1958 Pollen-analysis in mineral soil: an interpretation of a podzol pollen-analysis by Dr. G. W. Dimbleby. *Flora* 146:321–327.
1967 Strip lynchets and soil erosion. *Antiquity* 161:66–67.
1975 *The history of the British Flora*, 2nd Edition. The University Press, Cambridge.
1978 *Fenland: its ancient past and uncertain future*. The University Press, Cambridge.

Godwin, H., and G. F. Mitchell
1938 Stratigraphy and development in two raised bogs near Tregaron, Cardiganshire. *New Phytologist* 37:425–454.
Greig, J.
1981 The investigations of a medieval barrel-latrine from Worcester. *Journal of Archaeological Science* 8:265–282.
1982 The interpretation of pollen spectra from urban archaeological deposits. In Environmental archaeology in the urban context, edited by A. R. Hall and H. K. Kenward. *Council for British Archaeology Research Report* 43:47–65.
Groenman-van Waateringe, W.
1978 The impact of Neolithic man on the landscape in the Netherlands. In The effect of man on the landscape: the Lowland Zone, edited by S. Limbrey and J. G. Evans. *Council for British Archaeology Research Report* 21:135–146.
1979 Palynological investigations of five German burial mounds. *Archaeo-Physika* 8:69–84.
Hall, A. R., H. K. Kenward, D. Williams, and J. Greig
1983 Environmental living conditions at two Anglo-Scandinavian sites. In *The Archaeology of York* 14/1: Council for British Archaeology, London.
Hall, S. A.
1981 Deteriorated pollen grains and the interpretation of Quaternary pollen diagrams. *Review of Palaeobotany and Palynology* 32:193–206.
Harrison Matthews, L.
1952 *British mammals*. Collins New Naturalist, London.
Havinga, A. J.
1963 A palynological investigation of soil profiles developed in cover sand. *Mededelingen von de Landbouw-hogeschool te Wageningen, Nederland* 63:1–92.
1964 Investigation into the differential corrosion susceptibility of pollen and spores. *Pollen et Spores* 6:621–635.
1967 Palynology and pollen preservation. *Review of Palaeobotany and Palynology* 2:81–98.
1971 An experimental investigation into the decay of pollen and spores in various soil types. In *Sporopollenin,* edited by J. Brooks, P. R. Grant, M. D. Muir, P. van Gijzel, and G. Shaw, pp. 446–479. Academic Press, London.
1974 Problems in the interpretation of pollen diagrams of mineral soils. *Geologie en Mijnbouw* 53:449–453.
Hayes, R. H.
1967 The chambered cairn and adjacent monuments on Great Ayton Moor, N. E. Yorks. *Scarborough and District Archaeological Society Research Report* 7.
Heim, J.
1979 Recherches paléobotaniques au site néolithique (Roessen) de la "Bosse de l'Tombe" a Givry. *Bulletin de la Société Royale Belge d'Anthropologie et de Préhistoire* 90:65–78.
Hodges, D.
1974 *The pollen loads of the honey bee*. Bee Research Association, London.
Holden, E. W., and Bradley, R. J.
1975 A late Neolithic site at Rackham. *Sussex Archaeological Collections* 113:85–103.
Hubbard, R. N. L. B.
1982 The environmental evidence from Swanscombe and its implications forpalaeolithic archaeology. In Archaeology in Kent to A.D. 1500, edited by P. Leach. *Council for British Archaeology Research Report* 48:3–7.
Iversen, J.
1941 The influence of prehistoric man on vegetation. *Danmarks Geologiske Undersøgelse,* Series 2(6):1–25.

1958 Pollanalytischer Nachweis des Reliktencharakters eines jütischen Linden-Mischwaldes. *Veröffentlichungen des Geobotanischen Institutes Rübel in Zürich* 33:137–144.
1964 Retrogressive vegetational succession in the Post-glacial. *Journal of Ecology* 52 (Supplement):59–70.

Jacobi, R. M.
1976 Britain inside and outside mesolithic Europe. *Proceedings of the Prehistoric Society* 42:67–84.
1980 Whitehill, Longmoor Inclosure. In Archaeology, *Hampshire Annual Report 1979*. Hampshire County Council, 1980.
1981 Whitehill, Longmoor Inclosure. In Archaeology, *Hampshire Annual Report 1980*. Hampshire County Council 1981.

Janssen, C. R.
1959 *Alnus* as a disturbing factor in pollen diagrams. *Acta Botanica Neerlandica* 8: 55–58.

Jashemski, W. F.
1979 *The gardens of Pompeii*. Caratzas, New York.

Jewell, P. A. (editor)
1963 *The experimental earthwork on Overton Down, Wiltshire, 1960*. British Association for the Advancement of Science, London.

Jewell, P. A., and G. W. Dimbleby
1966 The experimental earthwork on Overton Down, Wiltshire, England: the first four years. *Proceedings of the Prehistoric Society* 32:313–342.

Jope, E. M., and H. M. Jope
1959 A hoard of fifteenth-century coins from Glenluce sand-dunes and their context. *Mediaeval Archaeology* 3:259–279.

Keef, P. A. M., J. J. Wymer, and G. W. Dimbleby
1965 A mesolithic site on Iping Common, Sussex, England. *Proceedings of the Prehistoric Society* 31:85–92.

Knights, B. A., C. Dickson, J. H. Dickson, and D. J. Breeze
1983 Evidence concerning the Roman military diet at Bearsden, Scotland, in the 2nd century A.D. *Journal of Archaeological Science* 10:139–152.

Körber-Grohne, U.
1967 *Feddersen Wierde. Die Ergebnisse der Ausgrabung der vorgeschichtlichen Wurt bei Bremerhaven in den Jahren 1955–1963*. Vol. 1. Franz Steiner Verlag GMBH, Wiesbaden.
1981 Crop husbandry and environmental change in the coastal area of the Feddersen Wierde, near Bremerhaven, Northwest Germany. In The Environment of Man: the Iron Age to the Anglo-Saxon Period, edited by M. Jones and G. Dimbleby. *British Archaeological Reports (British Series)* 87:287–307.

Lambrick, G., and M. Robinson
1979 Iron Age and Roman riverside settlements at Farmoor, Oxfordshire. *Oxfordshire Archaeological Unit Report* 2.

Lang, G.
1962 Vegetationsgeschichtliche Untersuchungen der Magdalénienstation an der Schussenquelle. *Veröffentlichungen des Geobotanischen Institutes Rübel in Zürich* 37:129–154.

Laudermilk, J. D., and R. A. Munz
1938 Plants in the dung of *Nothrotherium* from Rampart and Muave Caves, Arizona. *Carnegie Institution of Washington, Publication* 487:271–281.

Lawton, R. M.
1959 A pollen analysis of the Lake Bangweulu peat deposits. *Northern Rhodesian Journal* 4:33–43.

Leroi-Gourhan, A.
1956 Notes sur l'analyse pollinique des sédiments quaternairs des grottes. *Congrés préhistorique de France, XVe session, Poitiers, 1955*:671–675.
1960 Résultats de l'analyse pollinique de la grotte d'Isturitz. *Bulletin de la société préhistorique française* 45:619–624.
1965 Les analyses polliniques sur les sediments des grottes. *Bulletin de l'association française pour l'étude du Quaternaire* 2:145–152.
1969 Pollen grains of Gramineae and Cerealia from Shanidar and Zawi Chemi. In *The Domestication and Exploitation of Plants and Animals,* edited by P. J. Ucko and G. W. Dimbleby, pp. 143–148. Duckworth, London.
1973 Les méthódes quantitatives d'étude des variations du climat au cours du Pléistocène. *Colloques internationaux du Centre National de la Recherche Scientifique* 219:61–66.
1975 The flowers found with Shanidar IV, a neanderthal burial in Iraq. *Science* 190:562–564.
Leroi-Gourhan, A., and J. Allain
1979 *Lascaux inconnu.* Editions du Centre National de la Recherche Scientifique, Paris.
Leroi-Gourhan, A., and M. Girard
1971 L'abri de la Cure à Baulmes (Suisse). Analyse pollinique. *Jahrbuch der schweizerischen Gesellschaft für Ur- und Frühgeschichte* 56:7–15.
Leroi-Gourhan, A., C. N. Mateesco, and E. Protopopesco-Pake
1967 Contribution a l'étude du climat de la station de Vadastra du paléolithique supérieur a la fin du Néolithique. *Bulletin de L'association Française pour L'étude du Quaternaire* 1967:271–279.
Louwe Kooijmans, L. P.
1980 Archaeology and coastal change in the Netherlands. In *Archaeology and coastal change,* edited by F. H. Thompson, pp. 106–133. Society of Antiquaries, London.
Mackney, D.
1961 A podzol development sequence in oakwoods and heath in central England. *Journal of Soil Science* 12:23–40.
MacPhail, R.
1981 Soil and botanical studies on the 'Dark Earth'. In The environment of man: the iron age to the Anglo-Saxon period, edited by M. Jones and G. Dimbleby. *British Archaeological Reports (British Series)* 87:309–331.
Manby, T. G.
1976 Excavation of the Kilham long barrow, East Riding of Yorkshire. *Proceedings of the Prehistoric Society* 42:111–159.
Martin, P. S.
1963 *The last 10,000 years.* The University of Arizona Press, Tucson.
Martin, P. S., and W. Byers
1965 Pollen and archaeology at Wetherill Mesa. *American Antiquity* 31:122–135.
Martin, P. S., and F. W. Sharrock
1964 Pollen analysis of prehistoric human feces; a new approach to ethnobotany. *American Antiquity* 30:168–180.
Martin, P. S., B. E. Sabels, and D. Shutler, Jr.
1961 Rampart Cave coprolites and ecology of the Shasta ground sloth. *American Journal of Science* 259:102–127.
Moore, P. D., and J. A. Webb
1978 *An illustrated guide to pollen analysis.* Hodder and Stoughton, London.
Mourer-Chauviré, C., and J. Renault-Miskovsky
1980 Le paléoenvironnement des chasseurs de Terra Amata (Nice, Alpes-Maritimes) au Pléistocène Moyen. La flore et la faune de grands mammifères. *Géobios* 13, fasc. 3:279–287.

Noel Hume, A.
1976 Historical archaeology in garden restoration. *Landscape Architecture,* May 1976:259–264.
Oldfield, F.
1960 The coastal mud-bed at Mouligna, Bidart, and the age of Asturian industry in the Pays Basque. *Pollen et Spores* 2:57–70.
1967 The palaeoecology of an early neolithic waterlogged site in northwestern England. *Review of Palaeobotany and Palynology* 4:67–70.
Orme, B. J.
1982 The use of radiocarbon dates from the Somerset Levels. *Somerset Levels Papers* 8:9–25.
Osborne, P. J.
1969 An insect fauna of Late Bronze Age date from Wilsford, Wiltshire. *Journal of Animal Ecology* 38:555–566.
Palmer, S., and G. W. Dimbleby
1979 A mesolithic habitation site on Winfrith Heath, Dorset. *Proceedings of the Dorset Natural History and Archaeological Society* 101:27–50.
Parent, R., and N. Planchais
1972 Nouvelles fouilles sur le site Tardenoisien de Montbani (Aisne):1964–1968. *Bulletin de la Société Préhistorique Française* 69:508–532.
Peck, R. M.
1973 Pollen budget studies in a small Yorkshire catchment. In *Quaternary Plant Ecology,* edited by H. J. B. Birks and R. G. West, pp. 43–60. Blackwell, Oxford.
Percival, J.
1980 *Living in the Past.* BBC, London.
Piggott, S., and G. Dimbleby
1955 A Bronze Age barrow on Turner's Puddle Heath. *Proceedings of the Dorset Natural History and Archaeological Society* 75:34–35.
Pilcher, J. R.
1969 Archaeology, palaeoecology and ^{14}C dating of the Beaghmore stone circle. *Ulster Journal of Archaeology* 32:73–91.
Pilcher, J. R., and A. G. Smith
1979 Palaeoecological investigations at Ballynagilly, a Neolithic and Bronze Age settlement in Co. Tyrone, N. Ireland. *Philosophical Transactions of the Royal Society* B286:345–369.
Pitts, M. W., and R. M. Jacobi
1979 Some aspects of change in flaked stone industries of the Mesolithic and Neolithic in Southern Britain. *Journal of Archaeological Science* 6:163–177.
Planchais, N.
1965 Etude palynologique. In L'ossuaire néolithique d'Éteauville, commune de Lutz-en-Dunois (Eure-et-Loir), edited by A. Nouel, M. Dauvois, G. Bailloud, R. Riquet, T. Poulain-Josein, N. Planchais and P. Horemans. *Bulletin de la Société Préhistorique Française* 62:633–636.
1972 Analyse pollinique et occupations humaines successives du site de Montbani, commune de Mont-Notre-Dame (Aisne). *Comptes rendus de l'Académie des Sciences, Paris* 274:185–187.
Rankine, W. F., W. M. Rankine, and G. W. Dimbleby
1960 Further excavations at a Mesolithic site at Oakhanger, Selborne, Hants. *Proceedings of the Prehistoric Society* 26:246–262.
Ray, A.
1959 The effect of earthworms on pollen distribution. *Journal of Oxford University Forestry Society* 7:16–21.

Renault-Miskovsky, J.
1971 Analyse pollinique des sediments post-glaciaires de l'abri de Chateauneuf-les-Martigues. *Bulletin de L'association Française pur L'étude du Quaternaire* 1971:33–42.
Rennie, P. J.
1958 Physical soil properties and the significance of soil pollen concentrations. *Geologiska Föreningens i Stockholm Förhandlingar* 80:340–342.
Reynolds, P. J.
1976 *Farming in the Iron Age* The University Press, Cambridge.
Robinson, M. A.
1978 A comparison between the effects of man on the environment of the first gravel terrace and flood plains of the Upper Thames Valley during the Iron Age and Roman Periods. In The effect of man on the landscape: the Lowland Zone, edited by S. Limbrey and J. G. Evans. *Council for British Archaeology Research Report* 21:35–43.
Robinson, M., and R. N. L. B. Hubbard
1977 The transport of pollen in the bracts of hulled cereals. *Journal of Archaeological Science* 4:197–199.
Rodden, R. J.
1962 Excavations at the early neolithic site at Nea Nikomedeia, Greek Macedonia (1961 season). *Proceedings of the Prehistoric Society* 28:267–288.
Romans, J. C. C., and L. Robertson
1975 Soils and archaeology in Scotland. In The effect of man on the landscape: the Highland Zone, edited by J. G. Evans, S. Limbrey and H. Cleere. *Council for British Archaeology Research Report* 11:37–39.
Scaife, R. G.
(in press) Pollen analysis. In Palynological analyses of West Heath Barrows VI, VIII and IX, Sussex, edited by PP. L. Drewett. *Sussex Archaeological Collections* 122.
Scaife, R. G., and P. J. Burrin
1983 Floodplain development in and the vegetation history of the Sussex High Weald and some archaeological implications. *Sussex Archaeological Collections* 121:1–10.
Schoenwetter, J.
1974 Pollen analysis of sediments from Salts Cave vestibule. In *Archaeology of the Mammoth Cave Area,* edited by P. J. Watson, pp. 97–105. Academic Press, New York.
Schweingruber, F. H.
1976 Pollenanalytische Untersuchungen über die Siedlungsverhältnisse der frühneolithischen Station Egolzwil 5. In Das jungsteinzeitliche Jäger-Bauerndorf von Egolzwil 5 im Wauwilermoos, edited by R. Wyss. *Archaeologische Forschungen.* Schweizerischen Landesmuseum, Zürich.
Shennan, S. J., and R. T. Schadla Hall
1981 The archaeology of Hampshire from the Palaeolithic to the Industrial Revolution. *Hampshire Field Club and Archaeological Society, Monograph* 1:8–25.
Shotton, F. W.
1978 Archaeological inferences from the study of alluvium in the lower Severn-Avon valleys. In The effects of man on the landscape: The Lowland Zone, edited by S. Limbrey and J. G. Evans. *Council for British Archaeology Research, Report* 21:27–32.
Simmons, I. G.
1969 Pollen diagrams from the North York Moors. *New Phytologist* 68:897–827.
Simmons, I. G., M. A. Atherden, P. R. Cundill, and R. L. Jones
1975 Inorganic layers in soligenous mires of the North York Moors. *Journal of Biogeography* 2:49–56.

Simmons, I. G., and G. W. Dimbleby
1974 The possible role of ivy (*Hedera helix* L.) in the mesolithic economy of Western Europe. *Journal of Archaeological Science* 1:291–296.
Simmons, I. G., and M. J. Tooley
1981 *The Environment in Prehistory*. Duckworth, London.
Smith, C.
1979 Fisherwick. *British Archaeological Reports. (British Series)* 61.
Smith, K., J. Coppen, G. J. Wainwright, and S. Beckett
1981 The Shaugh Moor Project: third report—settlement and environmental investigations. *Proceedings of the Prehistoric Society* 47:205–273.
Thomas, N., and C. Thomas
1955 Excavations at Snail Down, Everleigh, 1953, 1955: an interim report. *Wiltshire Archaeological and Natural History Magazine* 56:127–148.
Thompson, M. W., and P. Ashbee
1958 Excavation of a barrow near the Hardy Monument, Black Down, Portesham, Dorset. *Proceedings of the Prehistoric Society* 23:124–136.
Tinsley, H. M., and R. T. Smith
1974 Ecological investigations at a Romano-British earthwork in the Yorkshire Pennines. *Yorkshire Archaeological Journal* 46:23–33.
Tomalin, D., and R. G. Scaife
1980 A neolithic flint assemblage and associated palynological sequence at Gatcombe, Isle of Wight. *Proceedings of the Hampshire Field Club and Archaeological Society* 36:25–33.
Troels-Smith, J.
1954 Pollenanalytische Untersuchungen zu einigen schweizerischen Pfahlbauproblemen. In *Das Pfahlbauproblem*, pp. 11–58, edited by W. U. Guyan. Birkhaüser Verlag, Basel.
1955 Pollenanalytische Untersuchungen zu einigen schweizerischen Pfahlbauproblem. *Monographien zur Ur- and Frühgeschichte der Schweiz* 11:11–56.
1960 The Muldbjerg dwelling place: an early neolithic archaeological site in the Aamosen bog, West-Zealand, Denmark. *The Smithsonian Report for 1959*:577–601.
Tubbs, C. R.
1968 *The New Forest: an ecological history*. David and Charles, London.
Tubbs, C. R., and G. W. Dimbleby
1965 Early agriculture in the New Forest. *Advancement of Science* 22:88–97.
Turner, J.
1962 The *Tilia* decline: an anthropogenic interpretation. *New Phytologist* 61:328–341.
1964 The anthropogenic factor in vegetation history. I. Tregaron and Whixall mosses. *New Phytologist* 63:73–90.
van Campo, M., and A. Leroi-Gourhan
1956 Note préliminaire à l'étude des pollens fossiles de différents niveaux des grottes d'Arcy-sur-Cure. *Bulletin du Muséum, 2nd series* 28:326–330.
van Zeist, W
1963 Het stuifmeelonderzoek van de grafheuvel Te Mol. *Archaeologica Belgica* 72: 20–22.
van Zeist, W., and W. A. Caspaire
1974 Niederwil, a palaeobotanical study of a Swiss lake shore dwelling. *Geologie en Mijnbouw* 53:415–428.
van Zinderen Bakker, E. M.
1969 The Pleistocene vegetation and climate of the basin. In *Kalambo Falls prehistoric site*, Vol. 1, edited by J. D. Clark, pp. 57–84. The University Press, Cambridge.

Vermeersch, P. M., A. V. Munaut, and E. Paulissen

1974 Fouilles d'un site du Tardenoisien final à Opglabbek-Ruiterskuil (Limbourg, Belge). *Quartär* 25:85–104.

Vermeersch, P., E. Paulissen, and A. V. Munaut

1973 Fouilles d'un site mesolithique à Opgrimbie (Limbourg, Belge) *Bulletin de la Société Royale Belge d'Anthropoligique et de Préhistoire* 84:97–152.

Vita-Finzi, C.

1978 *Archaeological sites in their setting.* Thames and Hudson, London.

Vita-Finzi, C., and G. W. Dimbleby

1971 Medieval pollen from Jordan. *Pollen et Spores* 13:415–420.

Vuorela, I.

1982 Pollen stratigraphy and chemical analyses of a mineral soil profile at a Corded Ware dwelling site in Southern Finland compared with those of local organic sediments. *PACT (Journal of the European Study Group on Physical, Chemical, and Mathematical Techniques Applied to Archaeology)* 7:175–193.

Vuorela, I., and M. Aalto

1982 Palaeobotanical investigations at a Neolithic dwelling site in southern Finland, with special reference to *Trapa natans*. *Annales Botanica Fennici* 19:81–92.

Walch, K. M., J. R. Rowley, and N. J. Norton

1970 Displacement of pollen grains by earthworms. *Pollen et Spores* 12:39–44.

Walker, D.

1957 A site at Stump Cross, near Grassington, Yorkshire, and the age of the Pennine microlithic industry. *Proceedings of the Prehistoric Society* 22:23–28.

Waterbolk, H. T.

1950 Palynological investigation of burial mounds. *Congrés International des Sciences Préhistoriques et Protohistoriques Zürich* 1950:130–133.

1954 *De praehistorische mens en zijn milieu.* Rijksuniversiteit te Groningen, Groningen.

1956 Pollen spectra from neolithic grave monuments in the northern Netherlands. *Palaeohistoria* 5:39–51.

Waterbolk, H. T., and W. van Zeist

1961 A Bronze Age sanctuary in the raised bog at Bargeroosterveld (Dr.) *Helinium* I:5–19.

1967 Preliminary report on the neolithic bog settlement of Niederwil. *Palaeohistoria* 12:559–580.

Welten, M.

1955 Untersuchungen über die neolithischen Siedlungsverhältnisse am Burgäschisee. In *Das Pfahlbauproblem*, pp. 61–88, edited by W. U. Guyan. Birkhaüser Verlag, Basel.

1958 Pollenanalytische Untersuchung alpiner Bodenprofile: historische Entwicklung des Bodens und säkulare Sukzession der örtlichen Pflanzengesellschaften. *Veröffentlichungen des Geobotanischen Institutes Rübel in Zürich* 33:253–274.

1967 Bemerkungen zur paläobotanischen Untersuchung von vorgeschichtlichen Feuchtbodenwohnplätzen und Ergänzungen zur pollenanalytischen Untersuchung von Burgäschisee-Süd. *Acta Bernensia* 2:9–20.

Whittington, G.

1983 A palynological investigation of a second millennium B.C. bank-system in the Black Moss of Achnacree. *Journal of Archaeological Science* 10:283–291.

INDEX

AUTHORS

GENERAL

PLACE NAMES

PLANT NAMES